Naked Politics

Naked Politics

Nudity, Political Action, and the Rhetoric of the Body

Brett Lunceford

LEXINGTON BOOKS
Lanham • Boulder • New York • Toronto • Plymouth, UK

Published by Lexington Books
A wholly owned subsidiary of The Rowman & Littlefield Publishing Group, Inc.
4501 Forbes Boulevard, Suite 200, Lanham, Maryland 20706
www.rowman.com

10 Thornbury Road, Plymouth PL6 7PP, United Kingdom

British Library Cataloguing in Publication Information Available

Library of Congress Cataloging-in-Publication Data
Lunceford, Brett.
Naked politics : nudity, political action, and the rhetoric of the body / Brett Lunceford.
p. cm.
Includes bibliographical references and index.
ISBN 978-0-7391-6709-0 (cloth : alk. paper)—ISBN 978-0-7391-7702-0 (electronic)
1. Nudity—Political aspects. 2. Human body—Political aspects. 3. Political participation. 4. Social action. 5. Political activists. I. Title.
BJ1500.N8L86 2012
306.4'613—dc23 2012013227

™ The paper used in this publication meets the minimum requirements of American National Standard for Information Sciences—Permanence of Paper for Printed Library Materials, ANSI/NISO Z39.48-1992.

Printed in the United States of America

For Rebecca.

Contents

Preface
It's Not Like You Haven't Seen a Naked Body Before

Everyone living on this planet has one thing in common. Each of us has a body. If one were to suddenly remove every article of clothing in existence, we would find that each of these bodies is remarkably similar in structure and form. Certainly there are almost infinite variations of the same theme, but when all is said and done, we are all naked underneath our clothes and all, with some notable exceptions, have similar parts. Yet there remains a strong taboo in Western society on seeing specific parts of the body that we all share in common. Were it not for this taboo, pornography would cease to exist as a multi-billion dollar industry. Indeed, one might even say that it is the taboo itself that makes the furtive viewing of another person's flesh exciting. After all, on seeing the other, one is faced with the realization that what he or she has witnessed was pretty much what was expected in the first place.

This is not to say that the body is not worth considering. A cursory glance at the history of art attests to our collective fascination with the body reaching back at least to the Venus of Willendorf. The human body is a thing of beauty, with an almost infinite variety of textures, colors, shapes, sounds, and scents to entrance us. Although we are seemingly all the same underneath our clothes, we realize that this sentiment is not entirely true. Otherwise, one would find little excitement in a lover's body. Disrobing with a lover for the first time is exciting because one immediately becomes privy to all of the nuances that make his or her body different—knowledge that was previously denied through the use of clothing. Such knowledge and intimacy can be intertwined, but this is not always the case. A physician asks the patient to disrobe to more accurately diagnose an ailment. One removes clothing to swim because bulky clothing can become waterlogged and become a hazard. One removes clothing to bathe. Parents change diapers and bathe children. In each of these instances, there is some instrumental value in removing clothing and these situations are commonly accepted in society.

What is less common—and less accepted—is nudity that is strategically employed as a mode of social and political action. This work examines the fundamental question of how the nude body can function rhetorically. I will confess at the outset that there is no easy answer to this question; if you were hoping for a definitive answer, you will be disappointed. Like all things rhetorical, the nude body functions in many different ways depending on who wields it. Sometimes context can be everything. The same words spoken at the appropriate moment from the mouth of a skilled orator can move a nation, but the same words uttered by a child may simply sound cute. Likewise, the unclothed body can be seen as erotic, dangerous, pathetic, disgusting, amusing, or simply naked, depending on who is naked, when, where, and why that display of nudity is taking place, and the skill with which it is employed. Like the endless varieties of the physical body, there are almost limitless ways to use nudity rhetorically. This work explores only a few of those possibilities.

In the interest of getting to the text itself, and because few read the preface anyway, I will conclude with the obligatory thanks that are due. I have had wonderful graduate assistants at the University of South Alabama who have helped me to gather material on the various forms of body protest, most notably Claire-Annick Melanson. I am grateful to Lexington Books for allowing me to reuse chapter 2, which was previously published in a slightly different form in *Arguments about Animal Ethics*, edited by Greg Goodale and Jason Edward Black. I have those editors to thank for helping me polish that chapter.

Finally, I have my family to thank. All of them have listened to me talk about these protests and actions for the last few years, but none has listened to more of it than my wife Rebecca. I am grateful for her support and after eleven years she is still my favorite person to see naked.

Chapter 1
Introduction: Naked in Public (For a Cause)

Georges Bataille observed that "the simple sexual act is different from eroticism; the former is found in animal life, whereas human life alone admits of an activity defined perhaps by a 'diabolical' aspect, aptly described by the word eroticism."[1] As with the distinction between sexual and erotic activity, there is likewise a difference between nudity as a natural state of being and nudity as erotic. Although the two are, in Western society at least, intertwined, this is not a book on erotic nudity. Rather, this work is concerned with how people have defined the nature of nudity itself and—more importantly—how nudity is employed as a means of social change. Of course nudity can be employed as a strategy to alter societal perceptions of erotic behavior, but the core focus of this work is on the politics of and through the unclothed body.

Public nudity as a form of protest has a long history reaching back at least as far as Lady Godiva's storied ride. Like all folklore, the story of Lady Godiva is a pastiche of fact and fiction with no authoritative account, but still it persists. According to legend, Lady Godiva rode through the streets of Coventry, England as a means of persuading her husband, Leofric, Earl of Mercia, to stop taxing his subjects so oppressively. He issued the challenge that if she would ride through the streets of Coventry naked, he would relent, believing that her piousness would prevent her from performing such an act. After issuing a proclamation that everyone should remain in their houses with the shutters and doors closed, she embarked on her ride. Because of this act, Coventry was freed from paying the tax. According to some versions of the legend, one person, a tailor named Tom, looked out and was struck blind, which spawned the notion of "peeping Tom."[2] Lady Godiva may well be the patron saint of politically motivated nudity.

But the history of public nudity goes back considerably further than Lady Godiva. One can trace it back to ancient Greece with Diogenes, the leader of the Cynics. According to the historian Diogenes Laertius, "It was his habit to do everything in public, the works of Demeter and Aphrodite alike"; once when he was "behaving indecently in public, he wished 'it were as easy to banish hunger by rubbing the belly.'"[3] Yet such sentiment can easily be understood in his assertion that the most beautiful thing in the world is "Freedom of speech."[4] Theodore Windt suggests that for the Cynics, "each act symbolically reinforced the commitment to a different way of living, even as it repudiated the conventional way in which men live."[5] For Diogenes, then, such displays in public were an affront to commonly accepted behaviors. For the ancient Greeks, male (but perhaps not female) nudity was seen as far less scandalous than now, but still, there are obvious norms which Diogenes was willing to violate.[6] His personal embodiment of the norms he wished to create makes this not only a philosophical statement, but a political one as well.

It seems that every generation has its cynics that strive to embody their ideology. The 1960s saw the rise of a counterculture of "flower children" that adopted a naturalist lifestyle in an effort to change social norms. Well known for their nudity, they used such tactics as a means of rejecting contemporary social values, including those surrounding sexuality.[7] But like Diogenes, those who engaged in these actions often did so in part because it was fun. Groups such as the Yippies demonstrated that the counterculture was not simply a somber response to a society gone astray.[8] The 1970s also saw the advent of the streaking craze, which Bill Kirkpatrick suggests seems trivial only in retrospect. Kirkpatrick notes that even conservative media outlets branded the practice as harmless fun, but that "contrary to the dominant narrative from 1974 forward, the story of streaking is not the story of a meaningless fad; it is the story of how streaking was turned into a meaningless fad through extensive discursive effort."[9] Yet this was mainly a function of the media. Sociological literature has challenged this interpretation, noting that "the behavior was meaningful and consequential to those enacting it."[10] However, this was not always the case. When asked why students streak, one woman at Florida State responded in a 1974 *Newsweek* article, "It's like playing bank robber . . . A guy jumps from the men's room without his clothes, runs across the campus and vanishes into a getaway car—just to see if he can get away with it. It's a challenge."[11] Yet for some university administrators, "Streaking represented the 'new morality' and was, therefore, perceived by many to be a challenge to traditional values and laws. In contrast, there were no moral or legal overtones to earlier fads like telephone-booth-stuffing or goldfish-swallowing."[12] As such, its utility as a means of social change cannot be overlooked.

More recently, there has been an explosion of nudity in political and protest action. For example, the animal rights organization AnimaNaturalis seems to have taken a page from People for the Ethical Treatment of Animals' (PETA) playbook with their campaign "Sin Piel" (without skin) employing public nudity en masse at a 2009 protest in Madrid, reminiscent of Spencer Tunick's work.[13] Environmental activists have also jumped naked into the fray, such as Dona Nieto (also known as La Tigresa), who recites poetry topless to loggers in an effort to have them see both her and the earth as vulnerable and sacred.[14] Ukrainian Feminist group FEMEN has consistently used bare-breasted women in their protest actions.[15] It seems that organizations and individuals all over the world have bought into the idea that exposing flesh is the best way to get attention for their causes, so much so that Philip Carr-Gomm argues that "the extent of its use in the English-speaking world now means that we have entered an era in which serious campaigns run the risk of losing their impact if the use of nakedness is not carefully managed."[16]

For some groups nudity itself is the key focus rather than simply a means to an end. For example, Bare Witness, a group that spells out words such as "peace" and "no war" on the ground with their naked bodies, states that they are "using the power and beauty of our bodies to send out a message of peace."[17] Other protest groups have followed similar tactics.[18] One woman in Greenwich Village protested the Iraq War by stripping naked to display "no war" and "stop the war," written on her body in red paint. She stated, "This is my only way to talk about my beliefs. It's a metaphor. When I am appearing naked, I have disarmed myself from any uniform because naked people, they never can make war."[19] Despite the protestor's incorrect assertion that naked people cannot make war, the naked body demonstrates a kind of vulnerability that lends itself well to anti-war protests.[20] Yet some have used nudity against ideological opponents who also use nudity. When protestors opposed to genetically modified food spelled out "No GMO" using their naked bodies, "about 300 Raelians spelled 'J'aime OGM' and 'I love GM' with their naked bodies in a Quebec field" in reply.[21]

In New York, a group of women, the "Topfree Seven," marched topless to assert their rights to topfreedom. This is a powerful rhetorical strategy for more than its shock value. Such an act embodies the very rights they wish to have or exercise, and action can often force the hand of the state. As Pundurs put it, "The women were interested in more than merely exercising their rights; they also declared their right to appear topfree in their community. They were not making a fashion statement; they were making a political statement."[22] It is this political stance that differentiates the Topfree Seven from other forms of exposure, such as nude sunbathing or nude dancing. Libby Adler argues that "a willing (or perhaps more accurately, an *eager*) audience is not all that distinguishes the topless dancers from the Topfree Seven. Santorelli and the rest of the Topfree Seven bared their breasts in a spirit of protest—not sexual availability."[23] Others, such as the Raelians, have also marched for topfreedom with protest actions such as their "National Go Topless Day."[24]

Disrobing in the name of a movement has also spilled over into the online world. Some have made a decidedly explicit connection between sexuality, nudity, and activism, such as "Fuck for Forest," which urges readers to become "erotic activists": "If you have self-made erotic photos or videos, please send it to us. Help create a greener and more sex positive planet."[25] The site had its genesis in a live music performance in which the founders literally had sex on stage to raise awareness concerning the plight of the rainforest. Although this was an interesting tactic, and the intersection of sexuality and nature speaks to what Sylvie Shaw would call the "ecoerotic," the site now seems more about seeing dreadlocked men and women having sex with other dreadlocked men and women.[26] Another group that claims to use nudity for activism is Super-

Beauty.Org.[27] This website has a link for "activism," which details injustices committed against women as a result of sexual behavior, such as a teacher who was fired for having nude photographs of herself on the web, or women in repressive nations who were sentenced for having sex. Even so, this site seems more a repository of erotica with an activism slant than a serious player in women's rights.

Other online groups have rejected the connection between nudity and eroticism, using exposure as a means of creating a sense of camaraderie and support. The website "The Shape of a Mother," for example, was created so "women of all ages, shapes, sizes and nationalities can share images of their bodies so it will no longer be secret. So we can finally see what women really look like sans airbrushes and plastic surgery. I think it would be nothing short of amazing if a few of our hearts are healed, or if we begin to cherish our new bodies which have done so much for the human race."[28] The site has many stories of pregnancy and birth and images of the mother's body during pregnancy and postpartum. Many of these photos have exposed breasts, bellies, or are completely nude, with some chronicling the transition from early pregnancy to several years after pregnancy. The women on the site write about their pre-pregnancy bodies, post-partum depression, their children, and the struggle to reclaim their bodies. Other women comment on the stories and photos, offering support and adding their own stories to validate each woman's experience. Many also comment on how beautiful the woman is. Their nudity is not used as titillation, but rather becomes the evidence for the narratives that they relate concerning the experience of motherhood. Their bodies have a story to tell—one that must be seen to be understood.

Social Movements and the Rhetorical Body

Sociologists John McCarthy and Mayer Zald define social movements as "a set of opinions and beliefs in a population which represents preferences for changing some elements of the social structure and/or reward distribution of a society."[29] But this does not take place strictly through rhetoric. Neil Smelser makes the connection between action and rhetoric clear when he defines collective behavior as "mobilization on the basis of a belief which redefines social action."[30] Indeed, McCarthy and Zald are likewise interested in the mobilization of individuals in social movement actions. They proposed the Resource Mobilization theory of social movements, which "examines the variety of resources that must be mobilized, the linkages of social movements to other groups, the dependence of movements upon external support for success, and the tactics used by authorities to control or incorporate movements."[31] A key resource that must be mobilized in any social movement is a mass of people. As Barbara Sutton observes, "political resistance involves, first and foremost, putting the material body in action to affect the course of society."[32] For a social movement to gain any traction it must mobilize bodies.

A mass of protesting bodies can be an awe-inspiring sight, mainly because they exist as more than simply bodies at a particular place and time. Rather, they have mobilized for a cause, to make a statement together. In this unity, the body becomes transformed into a part of the movement action. D. R. Koukal argues that protesting bodies are "a collective will concretized, an intersubjective mass animated by a common purpose that fills a public space and obstinately makes their shared demand."[33] Koukal further explains that "Dissenting bodies may orient themselves with other dissenting bodies, and acting in concert from a sense of shared political will, constitute themselves into a dissenting body en masse, which may vary in number and extent. As the dissenting body en masse becomes increasingly synthesized and dynamic, it becomes increasingly meaningful."[34]

In protest, the body becomes something more than a representative individual that desires change; it becomes a site of resistance. As Abby Peterson argues, "In the struggle against existing structures of power, the body is the most primary interface of powers of resistance. However, the body is more than an interface of the power of resistance, the militant body is a power of resistance."[35] Sutton likewise suggests that "activists' bodies convey political ideas, demand changes, and exert pressure. At the same time, activists project images about and experience their own bodily powers and capacities."[36] This resistance can be seen as threatening to those who are opposed by this mass because simply assembling can be an act of defiance: "Even standing mute in a relatively static state, the dissenting body en masse gently roils in animation, appearing to breathe. Simply standing silently, unified and obstinate, it has transformed the meaning of the place it inhabits. Its simple presence in this place is an *assertion* that takes the form of a silent *contention* with the political institutions it stands against."[37]

John Dewey reminds us that "ideas belong to human beings who have bodies."[38] These ideas are continually open for revision in the public sphere and one way that this takes place is through protest. Indeed, sociologist Verta Taylor suggests that in a utopian society "social protest would be routine."[39] If individuals are to make their desires known, they must act in public. This is the heart of democratic practice; as Stephen Frantzich explains, "democratic governments are based on the principle of responsiveness to the needs and desires *of those who make their needs and desires known*. . . . Just as democracy is not a spectator sport, it is also clear that one is unlikely to win if one does not play."[40] This process is the province of rhetoric, which Aristotle defined as "an ability, in each [particular] case, to see the available means of persuasion."[41] In the case of protest, the body is one such means of persuasion.

In considering how the body functions rhetorically, we must question the nature of the body itself. Elizabeth Grosz argues for a conception of corporeality that refuses "singular models, models which are based on one type of body as the norm by which all others are judged. There is no one mode that is capable of representing the 'human' in all of its richness and variability."[42] Following Grosz, Ruth Barcan argues that "if there is no singular 'body,' it follows that there is no singular 'nude.'"[43] However, specific attributes and differences invite different understandings of specific bodies. Thus, if one is to distill the rhetorical force of body display, he or she must account for differences among bodies. This is not to say that such judgments should be normative, advocating for a particular type of body, but differences matter. Perhaps the multiplicity of bodies is what makes their nudity so compelling. To witness another's naked body is to be confronted with a vision that is at once strange and familiar. As a rhetorician, I am more interested in how bodies are put into action and the ways in which one is invited to understand their exposure than in whether they measure up to a particular standard—cultural, moral, or otherwise.

It may be that the only time that a body is not rhetorical is when one is sleeping or dead. Yet even in these moments the body can be used rhetorically. Andy Warhol filmed John Giorno sleeping and produced the film *Sleep* "just when he thought sleep itself was becoming obsolete (a side effect of his own amphetamine use, as well as that of his Factory regulars)."[44] It can be argued that this was more of an aesthetic statement than a rhetorical one, but there are certainly cases in which the sleeping body can be rhetorical, such as an individual sleeping outdoors or in some unconventional location. One may even use sleep as a means of communicating disdain or boredom.

The case of Emmett Till is a striking example of the dead body used in a rhetorical fashion, as the images of his bloated corpse shocked a nation unaccustomed to the brutal culture of lynching in the Deep South. The *New York Times* noted that "if the lynching of Emmett Till was, as the historian David Halberstam called it, the first great media event of the civil rights movement, it became so largely because of the photographs of that monstrous face."[45] Indeed, Christine Harold and Kevin DeLuca argue that "the 'shock waves' produced by the corpse of Emmett Till in the mid-1950s continue to reverberate powerfully in the memory of those exposed to it. We suggest that they do so, in large part, because of the forcefulness of the image of the human body in peril. The dissemination and reception of this image—of the severely mutilated face of a child—illustrates the rhetorical and political force of images in general and of the body specifically."[46]

While conscious, people use their bodies to portray a desired presentation of self.[47] Movement, proximity, and other nonverbal behavior communicate volumes about the individual, and although some of these actions are subconscious, others are not. The clothing one chooses to wear, for example, has communicative value.[48] But the body itself, independent of its apparel and motions, can be the site and substance of rhetoric. For example, in the film *Super Size Me*, Morgan Spurlock demonstrates the unhealthy effects of fast food by documenting the changes in his body while eating nothing but McDonalds for thirty days. Ross Singer calls this strategy "*spectacular self-subjugation*—an activist tactic by which the body is given up temporarily to an exploitive system as a means of staging carnivalesque resistance against that system before a mainstream audience."[49] Spurlock's physical deterioration put him at severe risk, yet "without this body rhetoric, the force of presence in the film's anti-corporate case would have diminished severely."[50]

To disrobe in order to make a statement is to engage in a form of body rhetoric that, although increasingly common, is still quite foreign to most people. It is not the disrobing that is unusual, but the rhetorical use of nudity. Not all nude bodies are the same; as Bataille observes, "Decency is a chance matter of time and place and standards vary continually, even with individuals."[51] There are times in which a certain degree of exposure is sanctioned. For example, when one is on the beach, men and women wear little more than what would be typically covered by underwear, yet if someone chose to walk down the street in his or her underwear, such behavior would be viewed with disapproval. In other words, context matters. One may disrobe for any number of reasons, but protest is rarely one of them. However, this can be a powerful strategy. As Leonard Michaud observes, "Just as [Cohen's 'Fuck the Draft'] emotive content is lost in a substitute expression such as 'Abolish the Draft,' taking one's clothes off must surely signify an intensity of feeling on a wholly different scale than simply showing dissatisfaction by carrying a sign."[52] Indeed, nudity can alter the message itself; Daniel Linz and colleagues found that dancers who wore pasties and g-strings communicated a different message from those who were totally nude.[53] Even so, the idea of using the unclothed body as a means of protest is taboo even for some nudist organizations. An article in the *Washington Post* reports that Mary Jane Kolassa, a spokeswoman for the American Association for Nude Recreation, states that the organization "espouses nude recreation in appropriate places, but does not use nudity 'for social commentary, or rebelliousness or an act of civil disobedience.'"[54]

What makes the nude body so interesting is the variety of meanings it can embody and the way it can intensify (or overshadow) one's message. In some cultures, public nudity has been used as a shaming mechanism not for the one disrobing, but for the observer. For example, the humble act of mooning has a long history. Thomas Fowler explains that this insult was used by Greeks during the Fourth Crusade and eighteenth century Cherokee warriors.[55] In Nigeria, women "exposed their naked bodies, and most particularly their vaginas, to impose on oil company male dealers 'social death' through ostracization, which was widely believed to lead to actual demise."[56] In other words, these acts of public nudity were more than symbolic acts; rather, such acts can be seen as actual threats of physical violence. Public nudity, however, is seen as a last resort and this contributes to its symbolic power. Community development scholars Terisa Turner and Leigh Brownhill note that in Africa "women who go naked implicitly state that they will get their demands met or die in the process of trying."[57] Whether functioning metaphorically or mystically, as in the case of the Nigerian women, the use of public nudity is a particularly interesting use of body rhetoric. As Carr-Gomm argues, "we are fascinated by the naked protest because of its inherently paradoxical nature. Human beings are at their most vulnerable when naked but, when engaged in a protest, are also strangely powerful."[58]

That the body can be used as a symbolic object seems clear, but it is unlike other symbols because the individual inhabiting that body has, presumably, some degree of agency in how that symbol is employed. Still, there are aspects of reading the body that fall outside of the individual's agency. As Grosz observes, bodies "cannot be understood as ahistorical, precultural, or natural objects in any simple way; they are not only inscribed, marked, engraved, by social pressures external to them but are the products, the direct effects, of the very constitution of nature itself."[59] Laura Verdi likewise argues that "the social perception of the body is always changing, diachronically and synchronically, according to the time and space where the body is inscribed, taking on different valences in the social."[60] As such, the question concerning just what the body as symbol is supposed to mean is difficult to answer. The body can be profitably viewed as a polysemic text in that it resists a consistent reading. This is especially the case when the body is unclothed. That a taboo is being flaunted is generally clear; to what ends the taboo is broken are sometimes less evident. But human beings are not only physical beings, but also political beings, thus the political is always intertwined with the body and how the body is clothed or disrobed is part of that being.

Rob Cover suggests that "the cultural system of frameworks and contexts of nakedness operate to control and police sexuality," but taboos on nakedness do more than reign in sexuality; controls on nakedness function as controls on the body itself—how one can appropriately use one's own body.[61] James Barker explains that concertive control, or the constant observation by those around us, can be much more coercive than traditional means of bureaucratic control in part because of its transparency.[62] This, of course, was the argument made by Michel Foucault in his discussion of panoptic control—there is little need to police the masses if one can get them to police themselves.[63] Although there are legal frameworks in place, the societal taboos against public nudity are sufficiently strong to dissuade those who otherwise may consider transgressing norms against body exposure. However, transgression can be a powerful means of persuasion, and there are many ways that individuals and groups have deployed the unclothed body as a means of social change, whether explicitly or unintentionally, despite social pressure (or perhaps because of it).

As a tool to garner attention, nudity is a wonderful tool for two reasons. First, there are evolutionary imperatives. Because nudity is connected to sexuality in American culture, disrobing protesters unknowingly draw on hundreds of thousands of years of instinct that connects sexual display with mating success. One study of nightclub interactions found that "females that wore combinations of clothing that revealed very little flesh or breast area attracted very few approaches by males and those that wore more revealing clothing and danced in a sexually suggestive [manner] attracted the most."[64] Another study takes an evolutionary approach, suggesting that in poor economic times, when economically desirable men are scarcer, women increase competition by appearing more sexually attractive and engaging in increased sexual display.[65] For females in particular, it seems, the display of flesh is a powerful strategy for attracting sexual attention from males.

Secondly, there are cultural imperatives. Americans have a love/hate relationship with nudity. It is both taboo and sought after. Pornography is a multi-billion dollar industry, yet anti-nudity laws often prevent women from sunbathing topless. As Rob Cover observes, "Twentieth-century western culture authorizes particular sites of nakedness, some of which are ostensibly and explicitly linked with the sexual, others which are not."[66] Nudity is acceptable when it is in the realm of fantasy, but not when it unexpectedly materializes before us in three dimensions. There are some instances in which social movement actions intersect with this realm of fantasy, as in the case of charity causes that employ the "nude calendar of everyday people" approach. However, this strategy has become common enough that it seems to have lost much of its original novelty. This, of course, is a danger for those who seek to titillate the public with seemingly transgressive nudity. The ubiquitous nature of nudity on the Internet, advertisements, cable television, movies, and even network television may be contributing to a decline in the effectiveness of the naked form to elicit a response in such cases. Even so, the presence of a naked physical body still evokes a visceral reaction.

Outline of Chapters

This book examines some specific examples of nudity as a means of social protest. These examples were selected because they provide an interesting cross-section of strategies and groups. There are certainly other groups that I could have examined, but I have tried to avoid groups with similar aims. I have also tried to provide a variety of types of protestors, ranging from the professionalized to the grassroots to individuals. Finally, I have avoided actions that are mainly artistic.[67] For example, Spencer Tunick's work can be used as protest, such as when he teamed up with Greenpeace for a photograph on the Aletsch Glacier in the Swiss Alps.[68] However, as Carr-Gomm observes of Tunick's photograph, "When a protest becomes a work of installation art, aesthetic appreciation replaces outrage and we are left with an admiration for the work which risks overriding any sense of concern we might have for the issue it is attempting to address."[69]

In chapter 2 I examine in detail PETA's use of chained, nude women in their protests of the circus and the "running of the nudes" in Pamplona, Spain. As the most commonly recognized social movement organization that employs public nudity for social and political ends, PETA is an important group to examine. In this chapter I argue that the use of nudity goes beyond mere titillation, although this is an important facet in gaining media attention. These acts allow the protestors to cast off their humanity and stand in solidarity with the animals that they wish to save. Nudity functions as a way for the protestors to both depict to others and experience for themselves the experience of the animals that they represent.

Chapter 3 focuses on "lactivists," or women who engage in politically motivated breastfeeding in public as a means to both normalize public breastfeeding and to protest the social and structural obstacles that they often face in feeding their children, despite widespread legal support for the practice. I examine both the physical protests and the online protests waged against Facebook in response to their policy of removing breastfeeding photos from the site. This chapter provides an example of how the space in which one chooses to protest can have serious implications for how the protest is framed in the media and perceived by the public.

In chapter 4, I examine the World Naked Bike Ride (WNBR), an annual event that takes place in cities across the world as a means of protesting both dependence on oil and the vulnerability that bicyclists experience in a world of automobiles. Because the WNBR is likely the largest nude protest action in the world, it is particularly worthy of examination. Like the lactivists, the WNBR is a decentralized movement, allowing individuals to form rides in their own cities. However, this decentralization allows for different interpretations of the action itself by the protestors.

Chapter 5 examines a contest that took place on CollegeHumor.com in which women were invited to write their favorite presidential candidate's name on their breasts and upload the image. I suggest that rather than serving merely as exhibitionism, many of the women used the contest to make serious political statements. As in chapter 3, this chapter considers how the medium influences the message and places particular constraints on the individual. This chapter also considers the interplay between the desires of the individual and the message that the individual wishes to convey through the power of the body.

The book concludes by considering how the nude body functions rhetorically in the public sphere. Although each of these actions has different aims, they all seem to share four communicative strategies: look at me; desire me; fear me; believe me. I argue that in the cases presented here, the strategy of nakedness itself can be seen as: natural, exposure, harmlessness, difference, and erotic power. These sometimes contradictory ways of being demonstrate the difficulty in crafting a coherent narrative concerning that most universal of possessions—our naked bodies.

Notes

1. Georges Bataille, *The Tears of Eros* (San Francisco: City Lights Books, 1989), 23.

2. For more on the legend of Lady Godiva, see Dorothy Appleton, "Lady Godiva," *Dawn*, no. 36 (2000): 12; H. R. Ellis Davidson, "The Legend of Lady Godiva," *Folklore*

80, no. 2 (1969): 107-21; E. Sidney Hartland, "Peeping Tom and Lady Godiva," *Folklore* 1, no. 2 (1890): 207-26.

3. Diogenes Laertius, *Lives of Eminent Philosophers*, trans. Robert Drew Hicks, vol. 2 (Cambridge, MA: Harvard University Press, 1931), vi, 69.

4. Laertius, *Lives*, vi, 69.

5. Theodore Otto Windt, Jr., "The Diatribe: Last Resort for Protest," *Quarterly Journal of Speech* 58 (1972): 5.

6. The exception to this is the Spartans, who seemed to be more accepting of both male and female nudity. Daniel Garrison writes, "Spartan girls wore scanty athletic clothing, but this proximity to female nudity was distasteful to the same Greeks who accepted full nudity in male athletes." Sarah Pomeroy also notes, "There is some doubt about whether the girls [of Sparta] exercised in the nude. However, Spartan art of the Archaic period portrays the nude female body, while the art of other Greek cities does not." See Daniel H. Garrison, *Sexual Culture in Ancient Greece* (Norman: University of Oklahoma Press, 2000), 176-77; Sarah B. Pomeroy, *Goddesses, Whores, Wives, and Slaves: Women in Classical Antiquity* (New York: Schocken Books, 1975), 36.

7. See David Bell and Ruth Holliday, "Naked as Nature Intended," *Body & Society* 6, no. 3 (2000): 132.

8. The work of Abbie Hoffman exemplifies this notion that dissent can be fun. See Abbie Hoffman, *The Best of Abbie Hoffman*, ed. Daniel Simon (New York: Four Walls Eight Windows, 1989).

9. Bill Kirkpatrick, "'It Beats Rocks and Tear Gas': Streaking and Cultural Politics in the Post-Vietnam Era," *Journal of Popular Culture* 43, no. 5 (2010): 1024.

10. B. E. Aguirre, E. L. Quarantelli, and Jorge L. Mendoza, "The Collective Behavior of Fads: The Characteristics, Effects, and Career of Streaking," *American Sociological Review* 53, no. 4 (1988): 581.

11. Cited in William A. Anderson, "The Social Organization and Social Control of a Fad: Streaking on a College Campus," *Journal of Contemporary Ethnography* 6, no. 2 (1977): 225.

12. Anderson, "Social Organization," 232.

13. AnimaNaturalis, "Protesta 'Sin Piel' En Madrid Diciembre De 2009," www.animanaturalis.org/v/548.

14. See Becky Trout, "Striptease to Save Trees," *American Forests* 107, no. 3 (2001): 19.

15. For a repository of their protest actions, see FEMEN, "Femen—The Official Web Site," www.femen.org/.

16. Philip Carr-Gomm, *A Brief History of Nakedness* (London: Reaktion, 2010), 130.

17. Bare Witness, "Bare Witness Peace," www.barewitness.org.

18. See Gwendolyn Richards, "Naked Protesters Plan Anti-War Display Whatever the Weather," *Vancouver Sun*, February 22, 2003, B4.

19. Anthony Ramirez, "War Protester in Village Takes Her Cue from Lady Godiva," *New York Times*, August 10, 2005, B4.

20. There is a long history of warriors fighting naked and, in many cases, being the fiercest in battle. See Michael P. Speidel, "Berserks: A History of Indo-European 'Mad Warriors,'" *Journal of World History* 13, no. 2 (2002): 253-90.

21. "With Friends Like These, Monsanto Needs No Enemies," USA Today, www.usatoday.com/tech/news/techpolicy/2003-08-06-raelians-biotech_x.htm. See also Nick

O'Malley and Jonathan Pearlman, "Raelian around the GM Flag," *Sydney Morning Herald*, August 8, 2003, 16.

22. Helen Pundurs, "Public Exposure of the Female Breast: Obscene and Immoral or Free and Equal," *In the Public Interest* 14 (1994-1995): 10. The case was *People v. Santorelli*, 80 N.Y.2d 875 (1992).

23. Libby S. Adler, "A Short Essay on the Baring of Breasts," *Harvard Women's Law Journal* 23 (2000): 222.

24. GoTopless.org, "National Go-Topless Day!" www.gotopless.org/.

25. Fuck for Forest, "About FFF," www.fuckforforest.com/en/about.html.

26. Sylvie Shaw, "Reclaiming the Ecoerotic: Celebrating the Body and the Earth," *Ecotheology: Journal of Religion, Nature & the Environment* 8, no. 1 (2003): 85-99.

27. SuperBeauty.Org, "Superbeauty.Org—Beauty Will Save the World!" www.superbeauty.org.

28. The Shape of a Mother, "The Shape of a Mother," www.theshapeofamother.com.

29. John D. McCarthy and Mayer N. Zald, "Resource Mobilization and Social Movements: A Partial Theory," *American Journal of Sociology* 82, no. 6 (1977): 1217-1218.

30. Neil J. Smelser, *Theory of Collective Behavior* (New York: Free Press, 1962), 8.

31. McCarthy and Zald, "Resource Mobilization and Social Movements," 1213.

32. Barbara Sutton, "*Poner El Cuerpo*: Women's Embodiment and Political Resistance in Argentina," *Latin American Politics & Society* 49, no. 3 (2007): 130.

33. D. R. Koukal, "Here I Stand: Mediated Bodies in Dissent," *MediaTropes* 2, no. 2 (2010): 109.

34. Koukal, "Here I Stand," 115.

35. Abby Peterson, *Contemporary Political Protest: Essays on Political Militancy* (Aldershot, England: Ashgate, 2001), 69.

36. Sutton, "*Poner El Cuerpo*," 141.

37. Koukal, "Here I Stand," 119.

38. John Dewey, *The Public and Its Problems* (Athens, OH: Swallow Press, 1991), 8.

39. Verta Taylor, "Mobilizing for Change in a Social Movement Society," *Contemporary Sociology* 29, no. 1 (2000): 223.

40. Stephen E. Frantzich, *Cyberage Politics 101: Mobility, Technology and Democracy* (New York: Peter Lang, 2002), 130.

41. Aristotle, *On Rhetoric: A Theory of Civic Discourse*, trans. George Alexander Kennedy (New York: Oxford University Press, 1991), 1355b30-32.

42. Elizabeth A. Grosz, *Volatile Bodies: Toward a Corporeal Feminism* (Crows Nest, NSW: Allen & Unwin, 1994), 22.

43. Ruth Barcan, *Nudity: A Cultural Anatomy* (New York: Berg, 2004), 71.

44. Graig Uhlin, "TV, Time, and the Films of Andy Warhol," *Cinema Journal* 49, no. 3 (2010): 4. For more on *Sleep*, see Branden W. Joseph, "The Play of Repetition: Andy Warhol's 'Sleep,'" *Grey Room*, no. 19 (2005): 22-53.

45. Shaila Dewan, "How Photos Became Icon of Civil Rights Movement," *New York Times*, August 28, 2005, 1.12.

46. Christine Harold and Kevin Michael DeLuca, "Behold the Corpse: Violent Images and the Case of Emmett Till," *Rhetoric & Public Affairs* 8, no. 2 (2005): 266. For more on the rhetorical use of the body of Emmett Till, see Courtney Baker, "Emmett Till,

Justice, and the Task of Recognition," *Journal of American Culture* 29, no. 2 (2006): 111-24; Glenda Dicker, "Let the People See What I've Seen: In Praise of Mamie Till," *Southern Quarterly* 45, no. 4 (2008): 152-54; Davis W. Houck, "Killing Emmett," *Rhetoric & Public Affairs* 8, no. 2 (2005): 225-62.

47. See Erving Goffman, *The Presentation of Self in Everyday Life* (Garden City, NY: Doubleday, 1959).

48. See Brett Lunceford, "Clothes Make the Person? Performing Gender through Fashion," *Communication Teacher* 24, no. 2 (2010): 63-68; Brett Lunceford, "Smeared Makeup and Stiletto Heels: Clothing, Sexuality, and the Walk of Shame," in *College Sex: Philosophy for Everyone: Philosophers with Benefits*, ed. Robert Stewart and Michael Bruce, 51-60 (Hoboken, NJ: Wiley-Blackwell, 2010).

49. Ross Singer, "Anti-Corporate Argument and the Spectacle of the Grotesque Rhetorical Body in Super Size Me," *Critical Studies in Media Communication* 28, no. 2 (2011): 136.

50. Singer, "Anti-Corporate Argument," 146.

51. Georges Bataille, *Erotism: Death & Sensuality*, trans. Mary Dalwood (San Francisco: City Lights Books, 1986), 92.

52. Leonard J. Michaud, "Decent Exposure: An Inquiry into the Constitutionality of State Statutes Proscribing Expressive Public Nudity Comment," *Loyola Law Review* 22 (1975): 1030.

53. Daniel Linz, Eva Blumenthal, Edward Donnerstein, Dale Kunkel, Bradley J. Shafer, and Allen Lichtenstein, "Testing Legal Assumptions Regarding the Effects of Dancer Nudity and Proximity to Patron on Erotic Expression," *Law and Human Behavior* 24, no. 5 (2000): 507-33.

54. Lisa Rathke, "Teens Test Naked City's Limits; Town Asks: Should There Be a Law?" *Washington Post*, September 24, 2006, A8.

55. Thomas L. Fowler, "Of Moons, Thongs, Holdings and Dicta: State v. Fly and the Rule of Law," *Campbell Law Review* 22, no. 2 (2000): 256.

56. Terisa E. Turner and Leigh S. Brownhill, "Why Women Are at War with Chevron: Nigerian Subsistence Struggles against the International Oil Industry," *Journal of Asian and African Studies* 39, no. 1-2 (2004): 67. Although not explicitly defined in the article, "male dealers" seem to be the local Nigerians that were complicit in oil company operations.

57. Turner and Brownhill, "Why Women Are at War With Chevron," 71. See also, Shirley Ardener, "A Note on Gender Iconography: The Vagina," in *The Cultural Construction of Sexuality*, ed. Patricia Caplan, 113-42 (London: Routledge, 1987); Phillips Stevens, Jr., "Women's Aggressive Use of Genital Power in Africa," *Transcultural Psychiatry* 43, no. 4 (2006): 592-99.

58. Carr-Gomm, *A Brief History of Nakedness*, 133.

59. Grosz, *Volatile Bodies*, x.

60. Laura Verdi, "The Symbolic Body and the Rhetoric of Power," *Social Analysis* 54, no. 2 (2010): 100.

61. Cover, "The Naked Subject," 67.

62. James R. Barker, "Tightening the Iron Cage: Concertive Control in Self-Managing Teams," *Administrative Science Quarterly* 38, no. 3 (1993): 408-37.

63. See Michel Foucault, *Discipline and Punish: The Birth of the Prison*, trans. Alan Sheridan, 2nd Vintage Books ed. (New York: Vintage Books, 1995).

64. Colin A. Hendrie, Helena D. Mannion, and Georgina K. Godfrey, "Evidence to Suggest That Nightclubs Function as Human Sexual Display Grounds," *Behaviour* 146, no. 10 (2009): 1344.

65. Russell A. Hill, Sophie Donovan, and Nicola F. Koyama, "Female Sexual Advertisement Reflects Resource Availability in Twentieth-Century UK Society," *Human Nature* 16, no. 3 (2005): 266-77.

66. Rob Cover, "The Naked Subject: Nudity, Context and Sexualization in Contemporary Culture," *Body & Society* 9, no. 3 (2003): 56.

67. I recognize that there are times in which nude art can be used as social commentary. However, I am more interested in the physical body itself and how it can be deployed for political ends. For an excellent example of artistic nudity used as social commentary and the controversy such artwork can spawn, see Ruth Barcan's discussion of the *Down by the Lake with Liz and Phil*, a sculpture that depicted Queen Elizabeth II and Prince Phillip nude on a park bench. Barcan, *Nudity*, 219-26.

68. "600 Strip Naked on Glacier in Global Warming Protest: Chilling Message from Wear-Nothing Activists to Do-Nothing Politicians," Greenpeace, www.greenpeace.org/international/en/news/features/naked-glacier-tunick-08182007/.

69. Carr-Gomm, *A Brief History of Nakedness*, 130.

Chapter 2
Chained Women and Running Nudes: PETA's Body Rhetoric

As discussed in the first chapter, social movement organizations are incorporating nudity into their protest actions and one organization that has made extensive use of this strategy is People for the Ethical Treatment of Animals (PETA). PETA has long been known for its provocative protest tactics and some of these campaigns have resulted in backlashes. For example, their "Holocaust on Your Plate" campaign, in which PETA juxtaposed images from Nazi death camps with animal mistreatment or farming with captions such as "to animals, all people are Nazis," was denounced by the Jewish community and the Anti-Defamation League.[1] Their "I'd Rather go Naked than Wear Fur" campaign features nude—but strategically covered—models, musicians, and actresses. The latter campaign has caused controversy among feminists who question whether the pornographic undertones do more to harm women than to save animals.[2] Moreover, some scholars suggest that some of PETA's tactics may be counterproductive; legal scholar Maneesha Deckha argues that "attention to objectification is important not simply because it may harm women, but because it undermines the posthumanist project in general that PETA seeks to advance."[3] Laura Hahn likewise argues that "PETA fails to change minds, because they are off-message when they titillate or challenge the media establishment through their campaign choices rather than staying focused on the goal of reducing cruelty to animals. Yet PETA succeeds in perpetuating PETA."[4] In spite of potential backlash, PETA has continued its relentless pursuit of media attention through sexually charged campaigns that are largely disseminated through the mass media, billboards, and advertisements.

PETA has incorporated nudity into its protest actions to the point where it seems to be an essential tactic in their repertoire of strategies. For example, it recently released a "State of the Union Undress" video that features a well-dressed young woman who engages in a strip tease as she delivers her speech in front of an American flag. The speech is intercut with applause and footage from the congressional floor during an actual Presidential State of the Union Address. In the speech she states, "We will use all legal means at our disposal in ways that will capture the public's imagination, spur debate about animal rights, and encourage people to set aside their busy schedules just for a moment to consider the staggering number of lives at stake. Often this will mean taking our clothes off."[5] As she continues to remove her clothing, she speaks about PETA's campaigns, such as those against KFC and furriers. She concludes her speech completely naked, stating, "In our tireless quest to save these animals from exploitation, we promise that we will work harder, we will shout louder, we will push further, and we will get nakeder than ever before."[6] The video then cuts to scenes of animal experimentation and factory farms.

It is clear that PETA has a strong affinity for using nudity in its protest activities and advertising. Although scholars have examined PETA's advertising campaigns,[7] campaigns that involve activists performing staged public protest actions, such as PETA's "Running of the Nudes" during the running of the bulls in Pamplona, Spain or public displays of women chained or caged to protest the treatment of circus animals, have received little scholarly attention. In these campaigns, PETA demonstrates its commitment to the promise to get "nakeder than ever before." This chapter examines PETA's display of chained, shackled women and the Running of the Nudes to explore how their use of public nudity functions rhetorically as a protest strategy. Although PETA's actions serve to attract media attention by providing media-friendly image events, my analysis suggests that these actions transcend mere shock value and spectacle. As critical theorist Guy Debord declares, "The spectacle is the acme of ideology."[8] These protest actions do more than simply embody the notion that sex sells. Rather, nude protest functions as a way for members of PETA to perform an ideology that equates humans and animals by providing a way for the protesters to become more animalistic by shedding their clothing and performing the animal role in the protest, e.g., playing the part of the bulls in the running of the bulls and the part of the caged animals in the circus. Moreover, these protests help to build cohesion and strengthen their collective identity by prescribing for the members the lengths to which they should go in defense of animal rights.

Chained Women

One tactic that PETA uses to protest the treatment of animals is the public display of chained or caged women. These protests are often associated with the arrival of a circus. A *St. Louis Post-Dispatch* story describes one such street protest:

> A few hours before the Ringling Bros. and Barnum & Bailey Circus opened a five-day run in St. Louis on Wednesday, a near-naked 18-year-old woman sat on a crowded downtown street corner, chains wrapped around her ankles, to protest what activists allege is the mistreatment of circus elephants. People stopped to stare and snap photos with cell phone cameras. It was 35 degrees in

> the sun, but a goose-bumped Amy Jannette said she'd sit there, "For as long as I can."[9]

But the final paragraphs of the article cast doubt on the efficacy of such a display, illustrating how the display of flesh can overshadow the message:

> On Wednesday, however, it was unclear how effective Jannette's protest was. Three men in dress shirts and ties walked by the near-naked woman and a large sign reading "Circus elephants: Shackled, lonely, beaten." Over lunch, they talked about what they had just seen. They wondered why she was out there.
>
> "You guys read the sign?" one of the men said.
>
> The other two just shook their heads.[10]

But what must this tactic accomplish in order to successful? After all, the news media covered the woman's protest and disseminated the message, even if those who were physically present ignored the message. Even if the message is ignored, these protest actions still seem to fulfill what rhetorician Richard Gregg refers to as the "ego function of the rhetoric of protest," in which "the primary appeal . . . is to the protestors themselves, who feel the need for psychological refurbishing and affirmation."[11] The woman can claim a moral victory as she demonstrates solidarity with the chained animals. She draws the attention of gawking onlookers with cell phone cameras and can reasonably believe that she is raising awareness of the plight of circus elephants. However, she asks nothing of the viewer. She functions not as argument but as statement and spectacle. Rhetorician J. Michael Hogan argues, "There is more to democratic persuasion—or at least there ought to be more—than making the news. Genuinely democratic discourse must not only attract media attention; it must empower citizens to act."[12] By adopting a strategy of spectacle, which, as Debord observes, "manifests itself as an enormous positivity, out of reach and beyond dispute,"[13] the protestor provides the concerned onlooker with no way to interact.

But spectacle may be enough in today's media saturated environment where sound bytes stand in for argument. Jannette's protest action is ideally suited to meet the constraints of sound byte news. This is likely by design; members of PETA have repeatedly demonstrated their media savvy. One example of this can be found in a media event in which PETA co-founder and president Ingrid Newkirk and another member of PETA lay in makeshift coffins in the middle of Times Square. The *New York Times* noted that, "Ms. Newkirk, 56, was technically not nude, in the nude sense of the word. She wore white underpants, white stockings so thin you could see goose bumps, and white, flower-shaped pasties over her nipples, though not for the square inch of warmth they provided. A veteran activist, she later explained that the media would not use a picture 'if it has a nipple.'"[14] PETA seemingly understands what it must do to both attract and keep media coverage. It realizes the limits of the news media and is careful to keep media events within prescribed boundaries. Rhetorical scholar Peter Simonson notes that what makes PETA and other animal rights organizations distinct from the long tradition of concern for animal welfare is the media context in which current organizations exist.[15] Simonson further explains that PETA has been adept at inserting itself into popular culture, including music, fashion, and their own print outlets aimed at the true believers. He writes, "Entertainment cultures and their media include plenty of distortions, but they are also the symbolic worlds that many of us happily inhabit, as advertisers know too well. In claiming these worlds for their moral crusade, PETA is reminding us that democratic politics needs its popular pleasures."[16] In short, PETA creates a sustained, all encompassing media presence that allows it to disseminate its message; integrating their political message with popular culture allows for the widest possible diffusion through the largest variety of outlets.

PETA seems to recognize that protest can function as entertainment and spectacle. If part of that spectacle involves a bit of flesh and eye candy, so be it. Journalist John Elvin writes, "Most of the naked protestors say there's nothing suggestive or erotic in their behavior; they are just doing what comes naturally, or they are playing the part of dead bodies that result from war. A glance at PETA's calendar Website provides assurance that animal-rights protestors are not so purist in their justifications."[17] PETA plays with the erotic charge that can come from seeing vulnerable, mostly naked women immobilized by chains or cages. As such, it protests the exploitation of animals through the exploitation of women. Rhetorician Edwin Black argues that rhetorical discourses imply an ideal auditor, for whom the discourse is designed, and this implied auditor can often be linked to a particular ideology.[18] Who then, is the ideal viewer of this spectacle? Two possibilities that stand out are those who would be aroused by such a scene and those who would be repulsed.

There can be no doubt that arousal is part of the spectacle; the spectacle could just as easily have been performed by a clothed protester. The choice to be partially disrobed is a conscious decision that adds little to a message such as: "Circus elephants: Shackled, lonely, beaten."[19] Concerning the fetishistic element of the caged woman, one allusion that cannot be lost on such a pop-culture literate organization as PETA is the connection to the "women in prison" movie genre.[20] But if arousal is the end result, does this actually help disseminate its message? With arousal as enticement, the ideal viewer would be drawn to the flesh, but as her/his eye lingered on the scene s/he would also take in the message. The viewer would then make the connection between caged women and caged animals. Yet, two of the men who witnessed Jannette's protest completely missed the point of the act, likely more interested in the display of her flesh than the symbolic act that was taking place. Perhaps one reason for this is the unpredictable quality of erotic emotions. Those who are aroused by the sight of a caged, vulnerable woman may be aroused by the fantasies invoked, yet these fantasies may range from a desire to save the woman from captivity (the preferred reading from the standpoint of PETA) to the more troubling desire to contain and control women through shackling, caging, and publicly humiliating them. Thus it is difficult to predict the likely response of an ideal viewer who requires arousal as incentive to observe the protest.

We can gain some insight into how nudity and sexual appeals work as persuasion by examining studies in advertising.[21] Advertising and marketing researchers Jessica Severn, George Belch, and Michael Belch found that "the use of sexual appeals . . . seemed to detract from the processing and retention of message arguments."[22] They found that brand recall was not affected, but recall of copy was hindered.[23] So it seems reasonable to concede that people may remember that the protester was involved with PETA, assuming that the logo is prominently displayed. However, in many of these actions, the woman holds a small paper with a slogan such as "Wild animals don't belong behind bars," with a small logo. Although in many cases the protesters are identified as PETA activists, in the case of Jannette's protest mentioned above, two of the three men that witnessed her protest did not even notice the sign and wondered what she was doing.[24] Therefore, even the brand can be overshadowed at times by the display of nudity. Moreover, marketing scholars Claire Sherman and Pascale Quester found that "adverts for products exhibiting greater congruence with nudity were more effective in creating positive attitudes and greater purchase intentions."[25] This may prove problematic for PETA because animal rights and erotic displays are not generally connected in the mind of the general public in the same way as perfume and sexuality.

Viewer response to nudity may also depend on the viewer. Business scholars Jaideep Sengupta and Darren Dahl found that "consumers who have favorable affective reactions toward sexual stimuli—such as those who possess intrinsically positive attitudes toward sex per se—should evaluate sex-based ads positively in comparison to nonsexual ads, whereas the reverse should be observed for those with relatively unfavorable affective reactions toward sexual stimuli."[26] The gender of the viewer also plays a role in how such advertisements are viewed. Marketing scholar Michael LaTour found that "in terms of ad response, a nude model wins the popularity contest with men but not significantly more than women's feelings about a semi-nude model."[27] Perhaps part of this can be traced to how men and women view the display of female nudity. Sociologist Beth Eck states, "When women view the seductive pose of the female nude, they do not believe she is 'coming on to' them. They know she is there to arouse men."[28] Thus women may have a cynical view toward such displays and may see them as cheapening the message through sexual appeals to men. Eck's research suggests that "for women, female nudes are objects to be studied, viewed, judged, and, above all, used as a comparison for the self."[29] As such, these displays are an invitation for women to consider their own appearances in comparison to the protester rather than an opportunity to reflect on the plight of animals. This may explain some of the consternation of women who decry the fact that the women that represent PETA are almost universally young, slim, and attractive.[30]

As evidenced by the advertising studies discussed above, PETA's nude campaigns are probably not effective in changing people's minds or even in getting audiences to engage in deliberations about PETA's messages. Yet PETA seems to buy into the conventional wisdom that sex sells, despite the possibility that this strategy may backfire. Perhaps this type of ideal viewer represents PETA's view of the media and what gets their attention, as well as the attention of the general public, which at times seems fixated on sex. Thus, this ideal viewer requires spectacle—preferably of a sexual nature—to even pay attention to PETA.

The other ideal viewer of this scene is one who would be repulsed by such a display. Such repulsion would likely come out of pity rather than disgust. The sight of a helpless, vulnerable woman spurs a desire to protect and release the chained woman. This action elicits a powerful rhetorical connection between the vulnerability and imprisoned existence of the animals and that of the protester. The display of women in cages is a pitiful sight that invites the viewer to transfer this pity to the animals that are likewise caged. The ideal viewer would note the similar state of the caged or shackled protester to that of the circus animal and make the mental connection that just as humans should not be caged or shackled for our amusement, animals likewise should be free. However, the extremity of the sight may impede the viewer from making this connection. The nakedness of the protester presents a level of exposure that moves the audience to discomfort rather than sympathy; the level of pity that the scene stimulates becomes too much to bear. As such, the desire to protect and release the woman is given only to the protester and this desire overwhelms the ability of the audience to feel the same emotions for the animals that the protesters represent.

But there is another audience here that is not external; perhaps these actions are directed not at the general population as a way to gain attention and persuade a reluctant public, but at the protesters themselves. Gregg notes that one aspect of the ego-function of rhetoric "has to do with *constituting* self-hood through expression; that is, with establishing, defining, and affirming one's self-hood as one engages in a rhetorical act."[31] Thus, by participating in these actions, protesters provide a model of what it means to be a member of PETA. They display a level of commitment that transcends that required in other social movement organizations. Debord argues that "the individual who in the service of the spectacle is placed in stardom's spotlight is in fact the opposite of an individual . . . In entering the spectacle as a model to be identified with, he renounces all autonomy in order himself to identify with the general law of obedience to the course of things."[32] PETA members are invited to experience the suffering and humiliating existence of animals in a personal way. By disrobing and placing themselves on display, they demonstrate that they are willing to go through discomfort, both physical and psychological, in order to draw attention to the plight of animals. This can be seen in slogans such as "we'd rather go naked than wear fur." It is not enough to simply believe it; a true believer must also demonstrate it by physically disrobing and participating in displays of public nudity. Although these displays may do little to alter the public mind, directed inwardly they define the level of commitment required of protesters and serve as a call to action for PETA members in general.

The Running of the Nudes

The running of the bulls takes place during the festival of San Fermín in Pamplona, Spain. This event was introduced to the English speaking world largely through Ernest Hemingway's book *The Sun Also Rises*.[33] The bulls are run through the town on the way to the arena in which the bullfights will take place. Spanish scholar Timothy Mitchell writes, "The 'running of the bulls' in Pamplona is only the most famous example of what takes place in hundreds of folk festivals throughout both Castilles, León, La Mancha, Extremadura, Aragon, and Valencia, where communities ostensibly pay homage to their patron saints by stampeding and harassing bovids before slaughtering them and partaking of their flesh."[34] The running of the bulls is now firmly entrenched in popular culture as a display of bravery or drunken stupidity, depending on how the imagery is employed.

PETA attempts to draw attention to the opposition to bullfighting by staging a mock running of the bulls using nude activists. Participants in the "Running of the Nudes" wear the traditional red scarves of those who run with the bulls, fake bull-horns, and perhaps underwear (many are not actually nude, but some of the women are topless). Many carry placards that read, "stop the bloody bullfight" or "bullfighting" with the circle and slash motif. Some have slogans written on their bodies. The website for the Running of the Nudes describes the event this way: "Just two days before the first bull run, hundreds of activists—most wearing little more than a red scarf and plastic horns—ran through the streets of Pamplona for the annual 'Running of the Nudes.' Compassionate and fun-loving people from around the world met in Pamplona for the run to show the city that it doesn't need to torture animals for tourism."[35] And tourism is a serious consideration in this protest. Journalist Hilliard Lackey notes that the protest has an economic impact that enables authorities to overlook the illegal aspect of the action: "Authorities, however, are putting forth some semblance of a prohibition against nudity. Nevertheless, tourist dollars keep getting in the way. Tourism is the lifeblood of this area of Spain. The Running of the Nudes uprising—while illegal and uncouth—brings in a few more millions."[36]

Although the website includes bullfighting facts and a store to buy merchandise, the focus is on the sexiness of the runners. For example, there is a "Sexiest runner spotlight" and a page describing some of the "sexy runners" that participate. In fact, much of the site seems focused on the sexiness of the participants. On the page describing some of the participants, PETA makes the following appeal for joining the action: "How about slipping into something more comfortable—like your birthday suit—and joining us for this festive, cheeky event which is full of babes, not bulls? Speaking of babes, check out the profiles and pics of these hot-to-trot hotties you could be partying with in Pamplona!"[37] The profiles feature photographs and a short biography, complete with "hobbies," "turn-ons," and "turn-offs," that resemble a centerfold profile. Unlike the chained women protest actions, the Running of the Nudes is a mixed gender event. However, PETA's focus is still heavily slanted toward the women; of the eight profiles listed, six are female and two are male, all are attractive, and all but one are twenty-nine years old or younger, with the oldest being a pretty thirty-three-year-old blond woman.

In the videos of the action, there is a carnivalesque atmosphere.[38] There are banners and balloons and plenty of flesh. The participants seem to simply mosey down the street, in contrast to the frantic pace of the running of the bulls. The prurient interest is also present; one reporter states, "One observer might have the best advice so far: 'Forget the bulls, chase the nudes.' Surely, it is better to run behind just one Coppertone señorita than to run in front of a hundred enraged bulls."[39] In news reports, PETA reinforces the idea that the Running of the Nudes is meant to be a fun event. A PETA member recruiting participants in Covington, England states, "Most people don't realise the bulls are running to the death in the other one. We've been recruiting people all month and our race is just a fun event."[40]

Protesters make a direct connection between their own nudity and the welfare of animals. One protester states, "I'm always proud to bare a little skin if it means helping animals keep theirs," echoing a common slogan from PETA.[41] However, others recognize that the message comes at personal expense. An eighteen-year-old woman who describes herself as "shy and retiring," states, "I am taking part in this to attract attention about the cruelty to the bulls. If I stood with a placard which said this was cruel nobody would take much notice. If I have to take my clothes off to get the message across, then so be it."[42] This raises a significant question concerning the action's effectiveness and the desired outcome. If passersby are unwilling to consider arguments concerning the cruelty of bullfights made by a clothed woman, are they any more apt to consider these arguments from a naked woman? What, then, is taking place and what kind of response is invited by such an act? The protest action seems geared to generate publicity and allows PETA access to the media through which it can explain why it is protesting the running of the bulls and the institution of bullfighting in general. In short, the action provides a platform that PETA may otherwise have lacked from which to disseminate their message.

PETA has seized on other opportunities to gain publicity by inviting government officials to participate in the Running of the Nudes. A short article in the *Columbus Dispatch* notes, "A biography of [Donald] Rumsfeld reveals that he once participated in the bull run. 'I invite you to tip the scales back a bit by coming to Spain to participate in the Running of the Nudes—also called the Human Race—an alternative to the cruel spectacle that occurs three days later,' PETA President Ingrid Newkirk wrote. No word on whether Rumsfeld plans to participate."[43] This provides an opportunity to gain some publicity while also providing the media with a lighthearted story, complete with amusing mental imagery.

In contrast to the chained women protests, the Running of the Nudes succeeds in making its statement, partly by getting the attention of the media in a comical manner. Once PETA has the attention of the media, it is able to make short statements concerning its objections to bullfighting. The Running of the Nudes is much less confrontational than other PETA actions. Moreover, the participation of both males and females, as well as the large number of participants in the protest, removes some of the charges of simply exploiting women for PETA's cause. In a mass protest such as the Running of the Nudes, it is more about a moving mass of flesh than the display of a few choice specimens of female nudity as is seen in the chained women displays.

The Running of the Nudes seems to function as a much more effective protest than the case of the caged women, at least in getting the general public to pay attention to the message, because the nudity does not overshadow the message itself. Perhaps this stems from the mode of participation. Rather than standing in symbolically for the circus animals that would be abused, the protesters in the Running of the Nudes stand in as humans in the place of both humans and bulls. Observers are able to see the protest for what it is—raising a serious concern while still poking fun at humanity. In the Running of the Nudes the spectacle enhances the message, where in the case of the caged women the spectacle becomes the message.

The Rhetorical Force of PETA's Nude Protest

How best to get one's message into the public arena is a persistent concern in social movement protest. Sociologists Pamela Oliver and Daniel Myers note that "the link between public events and the public sphere is the mass media," and observe that "usually, a major goal of a public event is to attract the attention of the mass media, for only through the mass media can people communicate beyond their immediate social setting."[44] One effective way to do this is to stage what rhetorician Kevin DeLuca calls "image events":

> [I]n today's televisual public sphere corporations and states (in the persons/bodies of politicians) stage spectacles (advertising and photo ops) certifying their status before the people/public *and* subaltern counterpublics participate through the performance of image events, employing the consequent publicity as a social medium through which to hold corporations and states accountable, help form public opinion, and constitute their own identities as subaltern counterpublics. Critique through spectacle, not critique versus spectacle.[45]

In the case of PETA, it is clear that spectacle plays an important role in constituting identity and gaining access to the mass media. Of course, the more extreme the action, the more likely it is to wind up on the evening news or in the newspapers.

However, the question remains of whether the costs of nude protest actions outweigh the benefits of gaining the attention of the media and an otherwise apathetic public. This is especially problematic when considering the use of the female body in these actions. Although scholars such as Deckha and Pace attempt to "identify moments of resistive agency"[46] in some of PETA's campaigns such as the "I'd Rather Go Naked than Wear Fur" campaign, such arguments do not seem to hold up quite as well in the case of the caged women. In her discussion of PETA's advertising, Deckha writes, "Through animalizing the bodies of consenting women (women as foxes), the ad attempts to make present the absent referent of those beings whose bodies are rendered completely abject and object in the constitution of human subjects."[47] In the case of the caged woman, the woman stands in for the absent animal, but fails to do so convincingly to the observer—she is still very much a naked, vulnerable, caged *woman*. This is problematic because, as Deckha observes, "Any campaign that relies on standard representations of women that associate them with and even reduce them to their bodies continues the very same logic of commodification and objectification that is used against animals."[48] She continues, "Reducing women to their bodies in a context of animality, whether by presenting them as sexualized 'bunnies' or 'foxes' or simply connecting their sexualized bodies to the idea of animals, solidifies the trajectory of thinghood. All the usual suspects of things, rather than persons, are still aligned: women, body, animals."[49]

Feminist critiques that PETA campaigns that employ the sexualized female body simply reinforce the objectification of women seem to hold true in the case of the caged women. However, this is less evident in the case of the Running of the Nudes, which seems to take a more gender-neutral approach. Perhaps one way that PETA can engage in erotically charged protest while avoiding the critique that they are merely objectifying women is to use both sexes in similar ways. Deckha notes that when PETA uses male nudity, it is comedic or nonsexualized, with the exception of black males and white males appearing with white women.[50] In the case of the Running of the Nudes, both sexes are used in ways that are simultaneously comedic and sexual, although as mentioned previously the emphasis remains tilted toward the sexual desirability of the female body.

On a practical level, PETA's strategy of nude protest seems to function well, if for different reasons, depending on the audience. The displays of chained women provide a powerful means of building a collective identity centered on the suffering and humiliation of animals. As a means of gaining access to the public consciousness, however, these actions seem too extreme and invite incredulous gawking rather than serious reflection on the issue at hand. Even if the mental image of the naked, chained woman remains (and it likely will), the reason for her bondage will likely be shortly forgotten, if the viewer ever even realized the cause. This raises questions concerning the efficacy of such extreme tactics. Research by persuasion and social influence scholars Joseph Scudder and Carol Bishop Mills examined the effects of shocking advocacy videos from PETA and found that it can raise the credibility of animal rights organizations and lower credibility of animal processing plants.[51] Yet they are silent on whether this actually translated into a desire to pay more attention to where their meat comes from or whether they would stop eating meat. Changes in belief do not necessarily translate into changes in behavior; as social theorist Jacques Ellul observed, "there is not necessarily any continuity between conviction and action."[52]

The Running of the Nudes seems to function as a more effective means to gain access to the public consciousness. The core factor contributing to this is the playful tenor of the protest action. This event is less confrontational both physically (it takes place before the annual running of the bulls rather than during the event) and emotionally (there is no confusion of desire). The protesters state that this is meant to be a fun event and the images seem to reinforce this; the naked protesters seem to be having a good time. This stands in stark contrast to the tactics used in the displays of chained women next to a sign reading "Shackled, lonely, beaten." Thus, despite the even more dire occasion of bullfighting—the bulls *will* be killed; the circus at least has incentive to keep their animals alive—PETA is able to gain access to the media and make statements concerning the inhumanity of bullfighting without appearing self-righteous or strident.

In the case of the Running of the Nudes, PETA is able to link the protest action to the event itself. This can function as the first step in redefining the protested event, which is an important task for any social movement. Those who have seen or heard of the Running of the Nudes will associate it with the running of the bulls; the two will be forever linked in the public mind. But sometimes a strategy can work too well. PETA has recently changed its strategy for opposing bullfighting in the following statement: "We have decided to hold more demonstrations outside Pamplona. First, we have no wish to bring more and more runners—and thus tourist money—to a city that continues to support an outdated and barbarically cruel form of entertainment. Second, we want demonstrations happening in capital cities across the world, in countries from which tourists embark on trips to Spain."[53] This was in response, ironically, to their success in generating attention to the protest action. Philip Carr-Gomm notes that "the Running of the Nudes event has become so successful that it has brought more tourism to Pamplona, thus ensuring that the bull run continues."[54] PETA has now teamed up with Spanish animal rights activists AnimaNaturalis to stage demonstrations of activists painted to look like injured, bloody bulls. What was fun and carnivalesque has now moved toward the serious, chained women style of protest, although still retaining its mixed gender nature.

In all of these actions PETA is able to disseminate a message mainly by getting the attention of the media. This is an indictment not only of the media but also of the general population. If observers are unwilling to listen to arguments concerning matters of public concern unless they are presented in the guise of spectacle, this does not bode well for the current state of the public sphere. Recall the protester who stated, "If I stood with a placard which said this was cruel nobody would take much notice. If I have to take my clothes off to get the message across, then so be it."[55] In their quest to gain access to the public sphere, members of PETA may see little alternative to participating in image events designed to shock or titillate the public. If the only means of gaining access to the public sphere is through spectacle, one can hardly be blamed for employing image events as a means of disseminating one's message.

Extreme protests grow out of extreme ideologies. Research by political scientists M. Kent Jennings and Ellen Andersen suggests that "the strength of ideological orientations . . . proved to be a strong corollary of attitudes about disruption. Even in a group of activists with decidedly liberal leanings, the intensity of those leanings pushed in the direction of support for confrontation."[56] PETA evidently sees its cause as a literal matter between life and death. As such, it is not enough to simply believe in its cause; PETA is an organization that asks its members not only to do something but to be something. To fully subscribe to PETA's ideology goes beyond opposition to animal cruelty; one must also become a vegan, eschew products made with animal products, oppose vivisection and animal experimentation, and oppose the use of animals in entertainment. Members of PETA share a common vision of what one should eat, wear, and believe. In this sense, eating and dressing are rhetorical acts that help define PETA's members. With this in mind, Jennings and Andersen's research suggests that PETA members would support extreme tactics of confrontation and protest.[57] As such, it comes as little surprise that they would engage in strategies such as nude protest, which is both taboo and confrontational.

Perhaps one reason for the power of this strategy lies in its resistance to cooptation; the strategy of nude protest will not likely be used by any opposition groups against them. Because of the strong taboo against public nudity in American culture, nudity is unlikely to be used by any but the most extreme, dedicated organizations. As such, social movement organizations that occupy more extreme ideological positions will likely maintain a relative monopoly on the tactic; it is unlikely to be used by those seeking to maintain the status quo. PETA can use this strategy as a badge that illustrates their bravery as well as their progressiveness. This is similar to rhetorician Haig Bosmajian's assessment of the use of obscenity by the New Left in the 1960s: "The shouting of the obscenities may be the youths' vehicle for demonstrating their sexual, social, and political liberation. . . . The liberation may in the end be illusory, but for the moment the youthful demonstrators see it as real."[58] Because public nudity transgresses societal norms, PETA is able to define it as a brave act, rather than mere exhibitionism or perversion, because their acts of public nudity are done in the name of a greater good, despite oppositional claims that such tactics are degrading.

Public nudity is unsettling both because it is unexpected and because of the strong taboo associated with it. In some ways it functions much like a diatribe. Theodore Windt explains that "the diatribe is to rhetoric what satire is to literature. Each attempts to reduce conventional beliefs to the ridiculous, thereby making those who support orthodoxy seem contemptible, hypocritical, or stupid. Each seeks laughter, but not for its own sake. Rather, laughter serves as a cleansing force to purge pre-conceptions about ideas, to redeem ignored causes, to deflate pomposity, to challenge conventional assumptions, to confront the human consequences of ideas and policies."[59] Poking fun at a revered custom, such as the running of the bulls, allows PETA to draw attention to its cause while associating it with another, more well-known event. Adding the element of nudity to the protest invites the viewer to laugh at the ridiculous scene, yet in doing so, the viewer may see the ridiculous nature of the running of the bulls as well. Raising the level of discomfort through nudity is a powerful way to make an argument because there is no way to argue with nudity; the observer is psychologically knocked off balance and assumes a defensive stance from the beginning. To witness such a spectacle is to giggle the uncomfortable laugh of one who sees that which should not be seen.

Part of the power in PETA's use of public nudity lies in the symbolism it evokes. In her discussion of Amerindians' use of animal parts, archeologist Chantal Conneller writes, "people take on the animal habitus in order to enter into a particular set of relationships with the world."[60] A similar impulse seems to be at work in the actions of PETA protestors. By removing their clothing, activists are symbolically becoming more animal-like. They have stripped away the civilizing garments of contemporary society and come closer to their "natural" state. Indeed, the very idea of nakedness is a human concern, not one of the animals. Ruth Barcan argues that "in our everyday way of thinking, animals are rarely thought of as being nude, even though you can buy blankets for your horse and coats for your dog. The use of the terms 'naked' or 'nude' for an animal is so unusual in everyday language that it is mostly found in satire or in humour."[61] By stripping off their clothing and chaining and caging themselves, they experience the world as naked, chained, caged animals experience it. DeLuca, in his discussion of Earth First! activists, writes, "Perched high in the Douglas fir, the protester sees the world from the tree's point of view and 'becomes' the tree. Rendered relatively immobile, his movements are limited to the swaying of the tree. The protester, like the tree, depends on nourishment to come to him. Finally, their fates are intertwined as the protester depends on the tree for support and shelter while the tree depends on the protester's presence to forestall the chainsaw."[62] In their discussion of becoming-animal, cultural theorists Gilles Deleuze and Félix Guattari write, "You do not become a barking molar dog, but by barking, if it is done with enough feeling, with enough necessity and composition, you emit a molecular dog."[63] PETA activists do not simply portray animals; they symbolically *become* animals.

Such symbolic acts of becoming animalistic constitute a powerful rhetorical strategy, albeit mostly self-directed. By complicating the performance of "humanity," PETA activists transgress expected modes of behavior and draw attention to themselves. This seems to be an end in itself in PETA's constant quest for media exposure. Indeed, Wendy Atkins-Sayre argues that "PETA's advertising campaigns allow the group to effectively blur the distinction between human and nonhuman animals, inviting viewers to rethink their own identities and, thus, their beliefs about animal rights."[64] But this strategy also functions as a way to distinguish PETA activists from the rest of humanity. Becoming animalistic enables them to not only demonstrate solidarity with their non-human allies, but also helps the protesters constitute themselves through the rhetorical act. Rhetorician Maurice Charland notes that, "ideology is material because subjects enact their ideology and reconstitute their material world in its image."[65] Through nude protest PETA enacts an ideology that considers animals to be equal to humans by becoming more animalistic, reinforcing its beliefs iconically. In this mode of being, it is better able to speak on behalf of those who have no voice.[66]

PETA's acts of nude protest, then, fulfill two rhetorical functions. First, it helps to galvanize the movement by providing its members opportunities to prove their devotion to the cause by stripping naked and displaying themselves in public. As members break this taboo they become symbolically more animalistic, which provides them with a different kind of viewpoint from which to protest on behalf of animal rights. It is easier to sympathize with animals when you have become more animalistic, if only for a short time. Second, the protests act as image events that draw media attention to their cause. PETA seems to come from the school of thought that any publicity is good publicity. Once PETA has the attention of the media, it is able to disseminate its message more effectively. Nudity is the loss leader that draws the media consumer into the marketplace of PETA's ideas. Yet Hahn argues that once PETA has gained the attention of the media, the use of sexual appeals may overshadow the actual desired message of reducing cruelty to animals: "The only reason for PETA to launch ineffective commercials, then, may be to maintain the support of the already-convinced. Corporations call this 'brand loyalty.' I call this bad activism."[67]

Body rhetoric can be a potent force in social movement rhetoric. Ethicist Baruch Spinoza stated, "No one knows in what way and by what means mind can move body,"[68] but it can equally be noted that it is difficult to know how the display of the body can move the mind. Deleuze observes, "A body affects other bodies, or is affected by other bodies."[69] The only thing for certain is that the display of the body will affect others physiologically and emotionally; the precise nature of that effect is, however, uncertain.[70] We respond to the body because it is often used in a symbolic manner. Cultural theorist and critic Kenneth Burke reminds us that humans "by nature respond to symbols," and that persuasion can also be directed toward the self, as well as others.[71] By mobilizing the bodies of PETA activists and encouraging them to become more animalistic through display of their naked bodies, PETA orchestrates an act to which viewers cannot help but respond while simultaneously directing persuasion toward the activists themselves as they perform—and reinforce their belief in—their ideology.

Although nude protest is becoming more common, it will likely remain the province of only the most liberal and extreme groups. However, it may be these groups that need such protest strategies the most. PETA's message of veganism and animal rights is incompatible with the status quo of American and British society. However, some question whether spectacle is the correct strategy for groups like PETA; Atkins-Sayer suggests that "in a contest of images in particular, the more conservative position gains preeminence as the most viable option. This is so because publics do not have to pull together and reconcile the status quo within their belief systems, as they do with messages of change expressed through image events."[72] In order to reach the public, PETA must find a way to package its message in a way that causes the general public to listen. PETA has learned well the power of spectacle in our media saturated society, recognizing that naked protesters will always draw both a curious crowd and an interested media that thrives on the reporting of image events. So long as we all keep looking at the spectacle, PETA will have great incentive to supply us with something to look at. But whether people will actually see what PETA wants them to see is another question altogether.

Notes

*A slightly different version of this chapter was originally published as Brett Lunceford, "PETA and the Rhetoric of Nude Protest," in *Arguments About Animal Ethics*, ed. Greg Goodale and Jason Edward Black, 97-112 (Lanham, MD: Lexington Books, 2010).

1. Cathryn Bailey, "We Are What We Eat: Feminist Vegetarianism and the Reproduction of Racial Identity," *Hypatia* 22, no. 2 (2007): 41; Brian Delevie and Isshaela Ingham, "Unconscionable or Communicable: The Transference of Holocaust Photography in Cyber Space," *Afterimage* 31, no. 6 (2004): 7.

2. See Lesli K. Pace, "Image Events and PETA's Anti Fur Campaign," *Women and Language* 28, no. 2 (2005): 33-41.

3. Maneesha Deckha, "Disturbing Images: PETA and the Feminist Ethics of Animal Advocacy," *Ethics & the Environment* 13, no. 2 (2008): 55.

4. Laura K. Hahn, "I'm Too Sexy for Your Movement: An Analysis of the Failure of the Animal Rights Movement to Promote Vegetarianism," in *Arguments About Animal Ethics*, ed. Greg Goodale and Jason Edward Black, 79-95 (Lanham, MD: Lexington Books, 2010), 89-90.

5. "PETA's 2007 State of the Union Undress," People for the Ethical Treatment of Animals, features.peta.org/stateoftheunion2007/f-stateoftheunion.asp.

6. "PETA's 2007 State of the Union."

7. Bailey, "We Are What We Eat"; Jill Howard Church, "A Whopper of a Campaign," *The Animals' Agenda* 21, no. 3 (2001): 10; Deckha, "Disturbing Images"; Delevie and Ingham, "Unconscionable or Communicable"; Hahn, "I'm Too Sexy for Your Movement"; Merle Hoffman, "Do Feminists Need to Liberate Animals, Too? Carol Adams Sees Feminism as a Visionary Philosophy That Includes Stewardship of the Earth," *On The Issues* 4, no. 2 (1995): 18-24, 54-56; Mark Hudis, "PETA Picks a Pic of Naked Pop Star," *MediaWeek* 5, no. 13 (1995): 44; Andrew Linzey, "Good Causes Do Not Need Exaggeration," *The Animals' Agenda* 20, no. 1 (2000): 24-25; Pace, "Image Events and PETA's Anti Fur"; Peter Simonson, "Social Noise and Segmented Rhythms: News, Entertainment, and Celebrity in the Crusade for Animal Rights," *Communication Review* 4, no. 3 (2001): 399-420.

8. Guy Debord, *The Society of the Spectacle*, trans. Donald Nicholson-Smith (New York: Zone Books, 1994), 151.

9. Todd C. Frankel, "Activists, Circus Take Wraps Off Feud over Animals' Treatment," *St. Louis Post-Dispatch*, November 17, 2005, A1.

10. Frankel, "Activists, Circus," A1.

11. Richard B. Gregg, "The Ego-Function of the Rhetoric of Protest," *Philosophy & Rhetoric* 4, no. 2 (1971): 74.

12. J. Michael Hogan, *The Nuclear Freeze Campaign: Rhetoric and Foreign Policy in the Telepolitical Age* (East Lansing: Michigan State University Press, 1994), 187.

13. Debord, *Society of the Spectacle*, 15.

14. Dan Barry, "In Times Sq. These Days, Naked Ladies Have Causes," *New York Times*, November 23, 2005, B2.

15. Simonson, "Social Noise," 402.

16. Simonson, "Social Noise," 416.

17. John Elvin, "Tough Times for Buff Protests," *Insight on the News* 19, no. 5 (2003): 17.

18. Edwin Black, "The Second Persona," *Quarterly Journal of Speech* 56 (1970): 112.

19. Frankel, "Activists, Circus," A1.

20. For more on the "women in prison" genre of movies see Suzanna Danuta Walters, "Caged Heat: The (R)Evolution of Women-in-Prison Films," in *Reel Knockouts: Violent Women in the Movies*, ed. Martha McCaughey and Neal King, 106-23 (Austin, TX: University of Texas Press, 2001).

21. In this section, I will review several studies concerning sex and/or nudity in advertising, but for a meta-analysis of this literature see Tom Reichert, "Sex in Advertising Research: A Review of Content, Effects, and Functions of Sexual Information in Consumer Advertising," *Annual Review of Sex Research* 13 (2002): 241-73.

22. Jessica Severn, George E. Belch, and Michael A. Belch, "The Effects of Sexual and Non-Sexual Advertising Appeals and Information Level on Cognitive Processing and Communication Effectiveness," *Journal of Advertising* 19, no. 1 (1990): 22.

23. Severn, Belch, and Belch, "Effects of Sexual and Non-Sexual Advertising," 21.

24. Frankel, "Activists, Circus," A1.

25. Claire Sherman and Pascale Quester, "The Influence of Product/Nudity Congruence on Advertising Effectiveness," *Journal of Promotion Management* 11, no. 2/3 (2005): 80.

26. Jaideep Sengupta and Darren W. Dahl, "Gender-Related Reactions to Gratuitous Sex Appeals in Advertising," *Journal of Consumer Psychology* 18, no. 1 (2008): 73. For the most part Americans in particular have a love/hate relationship with nudity and sexuality. In a study that used advertisements of a product, a woman in a swimsuit, a topless model, and a nude model, Sid Dudley found that "the advertisement featuring the nude model was viewed as the most offensive, immoral, unethical, and exploitative." Such an attitude would certainly affect how the observer views the display. Sid C. Dudley, "Consumer Attitudes toward Nudity in Advertising," *Journal of Marketing Theory and Practice* 7, no. 4 (1999): 92.

27. Michael S. LaTour, "Female Nudity in Print Advertising: An Analysis of Gender Differences in Arousal and Ad Response," *Psychology & Marketing* 7, no. 1 (1990): 78.

28. Beth A. Eck, "Men Are Much Harder: Gendered Viewing of Nude Images," *Gender & Society* 17, no. 5 (2003): 706. For more on the impact of gender on the viewing of sexual advertising, see Richard Elliott, Abigail Jones, Andrew Benfield, and Matt Barlow, "Overt Sexuality in Advertising: A Discourse Analysis of Gender Responses," *Journal of Consumer Policy* 18, no. 2,3 (1995): 187-217; Marilyn Y. Jones, Andrea J. S. Stanaland, and Betsy D. Gelb, "Beefcake and Cheesecake: Insights for Advertisers," *Journal of Advertising* 27, no. 2 (1998): 33-51.

29. Eck, "Men Are Much Harder," 698.

30. See Nikki Craft, "PETA: Where Only Women Are Treated Like Meat," www.nostatusquo.com/ACLU/PETA/peta.html; kitchenMage, "Dear PETA, Women Are Animals Too," blog.kitchenmage.com/2007/10/dear-peta-women.html; Pace, "Image Events and PETA's Anti Fur."

31. Gregg, "Ego-Function," 74.

32. Debord, *Society of the Spectacle*, 39.

33. See Donnovan Andrews, "Bullish Steps," *Black Enterprise* May 2002, 114.

34. Timothy J. Mitchell, "Bullfighting: The Ritual Origin of Scholarly Myths," *Journal of American Folklore* 99 (1986): 405.

35. People for the Ethical Treatment of Animals, "Running of the Nudes: Out with the Old-in with the Nude!" http://www.runningofthenudes.com.

36. Hilliard Lackey, "Running of the Bulls and Running of the Nudes, Intriguing Options," *Mississippi Link* (Jackson, MS), July 6-July 12, 2006, B6.

37. People for the Ethical Treatment of Animals, "Running of the Nudes: Out with the Old-in with the Nude! // Next Year's Race // Look Who's Running," www.runningofthenudes.com/look_whos_running.asp.

38. I use the term "carnival" much as literary critic Mikhail Bakhtin does: "Because of their obvious sensuous character and their strong element of play, carnival images closely resemble certain artistic forms, namely the spectacle." Mikhail M. Bakhtin, *Rabelais and His World*, trans. Hélène Iswolsky (Bloomington: Indiana University Press, 1984), 7.

39. Lackey, "Running of the Bulls and Running of the Nudes," B6.

40. "All the Nudes That's Fit to Print," *Coventry Evening Telegraph*, June 23, 2006, 30.

41. Bill Radford, "Bullfight Protester Goes Bare; Woman with Local Ties Works for PETA," *The Gazette* (Colorado Springs, CO), June 25, 2005, Life3.

42. "Emma to Bare All for Naked Truth over Cruelty to Bulls," *Journal* (Newcastle-upon-Tyne, UK), June 19, 2006, 7.

43. Andrew Sterling, "Director Offers Version of the Hole Truth," *Columbus Dispatch*, May 20, 2005, E12.

44. Pamela E. Oliver and Daniel J. Myers, "How Events Enter the Public Sphere: Conflict, Location, and Sponsorship in Local Newspaper Coverage of Public Events," *American Journal of Sociology* 105, no. 1 (1999): 38, 39.

45. Kevin Michael DeLuca, *Image Politics: The New Rhetoric of Environmental Activism* (Mahwah, NJ: Lawrence Erlbaum Associates, 1999), 21-22.

46. Deckha, "Disturbing Images," 60. Pace also notes that "PETA's ads work simultaneously with and against the system." Pace, "Image Events and PETA's Anti Fur," 38.

47. Deckha, "Disturbing Images," 59.

48. Deckha, "Disturbing Images," 55.

49. Deckha, "Disturbing Images," 55

50. Deckha, "Disturbing Images," 51.

51. Joseph N. Scudder and Carol Bishop Mills, "The Credibility of Shock Advocacy: Animal Rights Attack Messages," *Public Relations Review* 35, no. 2 (2009): 162-64.

52. Jacques Ellul, *Propaganda: The Formation of Men's Attitudes*, trans. Konrad Kellen and Jean Lerner (New York: Knopf, 1965), 28.

53. People for the Ethical Treatment of Animals, "PETA Holds Nude Conferences in New Plan to Save Bulls," www.runningofthenudes.com/nude_conferences.asp.

54. Philip Carr-Gomm, *A Brief History of Nakedness* (London: Reaktion, 2010), 106-7.

55. "Emma to Bare All," 7.

56. M. Kent Jennings and Ellen Ann Andersen, "Support for Confrontational Tactics among AIDS Activists: A Study of Intra-Movement Divisions," *American Journal of Political Science* 40, no. 2 (1996): 331.

57. Jennings and Andersen, "Support for Confrontational Tactics," 331.

58. Haig A. Bosmajian, "Obscenity and Protest," *Today's Speech* 18 (1970): 11.

59. Theodore Otto Windt, Jr., "The Diatribe: Last Resort for Protest," *Quarterly Journal of Speech* 58 (1972): 8.

60. Chantal Conneller, "Becoming Deer. Corporeal Transformations at Star Carr," *Archaeological Dialogues* 11, no. 1 (2004): 43.

61. Ruth Barcan, *Nudity: A Cultural Anatomy* (New York: Berg, 2004), 65.

62. Kevin Michael DeLuca, "Unruly Arguments: The Body Rhetoric of Earth First!, ACT UP, and Queer Nation," *Argumentation & Advocacy* 36, no. 1 (1999): 13.

63. Gilles Deleuze and Félix Guattari, *A Thousand Plateaus: Capitalism and Schizophrenia*, trans. Brian Massumi (Minneapolis, MN: University of Minnesota Press, 1987), 275.

64. Wendy Atkins-Sayre, "Articulating Identity: People for the Ethical Treatment of Animals and the Animal/Human Divide," *Western Journal of Communication* 74, no. 3 (2010): 311.

65. Maurice Charland, "Constitutive Rhetoric: The Case of the *Peuple Quebecois*," *Quarterly Journal of Speech* 73, no. 2 (1987): 143.

66. For more on the idea of PETA serving as a voice in defense of (and in place of) animals, see Jason Edward Black, "Extending the Rights of Personhood, Voice, and Life to Sensate Others: A Homology of Right to Life and Animal Rights Rhetoric," *Communication Quarterly* 51, no. 3 (2003): 323, 327.

67. Hahn, "I'm Too Sexy for Your Movement," 94.

68. Baruch Spinoza, *The Ethics; Treatise on the Emendation of the Intellect; Selected Letters*, ed. Seymour Feldman, trans. Samuel Shirley, 2nd ed. (Indianapolis, IN: Hackett, 1992), 105.

69. Gilles Deleuze, "Ethology: Spinoza and Us," in *The Body: A Reader*, ed. Mariam Fraser and Monica Greco (New York: Routledge, 2005), 58.

70. Although I could find no studies that measured physiological response to the unexpected viewing of a live, nude body, in Western culture it is difficult to separate nudity from the erotic. Thus, I make this claim based on studies conducted by cognitive scientists and psychologists that suggest that erotic stimuli have a variety of possible effects on the viewer. Some of these researchers also note that intervening variables can affect how one responds to erotic stimuli. See Giulia Buodo, Michela Sarlo, and Daniela Palomba, "Attentional Resources Measured by Reaction Times Highlight Differences within Pleasant and Unpleasant, High Arousing Stimuli," *Motivation & Emotion* 26, no. 2 (2002): 123-38; C. Jiao, B. Turman, P. Weerakoon, and P. Knight, "Alterations in Grip Strength During Male Sexual Arousal," *International Journal of Impotence Research* 18, no. 2 (2006): 206-9; S. W. Kim, D. W. Sohn, Y. H. Cho, W. S. Yang, K. U. Lee, R. Juh, K. J. Ahn, et al., "Brain Activation by Visual Erotic Stimuli in Healthy Middle Aged Males," *International Journal of Impotence Research* 18, no. 5 (2006): 452-57; Cindy M. Meston and Boris B. Gorzalka, "Differential Effects of Sympathetic Activation on Sexual Arousal in Sexually Dysfunctional and Functional Women," *Journal of Abnormal Psychology* 105, no. 4 (1996): 582-91; Ion G. Motofel and David L. Rowland, "The Physiological

Basis of Human Sexual Arousal: Neuroendocrine Sexual Asymmetry," *International Journal of Andrology* 28, no. 2 (2005): 78-87; Harald T. Schupp, Bruce N. Cuthbert, Margaret M. Bradley, Charles H. Hilman, Alfons O. Hamm, and Peter J. Lang, "Brain Processes in Emotional Perception: Motivated Attention," *Cognition & Emotion* 18, no. 5 (2004): 593-611; Mark Spiering, Walter Everaerd, and Erick Janssen, "Priming the Sexual System: Implicit Versus Explicit Activation," *Journal of Sex Research* 40, no. 2 (2003): 134-45.

71. Kenneth Burke, *A Rhetoric of Motives* (New York: Prentice-Hall, 1950), 43, 38.

72. Wendy Atkins-Sayre, "Protection from 'Animal Rights Lunatics': The Center for Consumer Freedom and Animal Rights Rhetoric," in *Arguments About Animal Ethics*, ed. Greg Goodale and Jason Edward Black, 147-62 (Lanham, MD: Lexington Books, 2010), 149.

Chapter 3
Weaponizing the Breast: Lactivism and Public Breastfeeding

When my wife became pregnant with our first child I became intimately aware of the staggering amount of preparation and care that pregnancy requires. This preparation begins even before conception as women are counseled to begin taking prenatal vitamins and adopt a healthier lifestyle. Once pregnant, the list of behaviors that will help ensure a healthy, well formed baby begins to encompass the future mother's entire life. She should sleep on her left side to ensure adequate blood flow to the fetus, avoid taking certain pain relievers and other medication, maintain a balanced diet, and avoid certain exercises and activities. For example, if she has a condition such as placenta previa, she may have to abstain from sexual intercourse. The prospective mother will do all of this and more in the interest of giving birth to a healthy, happy baby girl or boy.

Of course this desire to nurture the child does not end at birth. An entire industry seems built around creating smart, strong children with advanced motor skills. Parents buy toys and products that are engineered to stimulate the baby's mind and body, such as Baby Einstein or Baby Mozart. In this quest, the parents will be aided by grandparents, friends, and advertisers who are all too happy to hawk their wares through commercials about diapers, toys, strollers, baby food, and medicine. Even strangers may offer advice or assistance when they see a crying baby. In short, society seems to facilitate the parents' desire for a healthy, well-adjusted child who will be a fully functioning member of tomorrow's society—with one exception: breastfeeding.

The medical community is in agreement on the fact that breast milk is the best food for a child, yet many in American society are quite uncomfortable when they encounter a woman nursing her child. When women choose to nurse in public spaces, they are often subjected to dirty looks by passersby or commanded by employees or security guards to either nurse in a restroom or leave the premises. Sometimes they are even denied services, as in the case of Emily Gillette who, along with her husband and child, was removed from a Freedom

Airlines flight before takeoff because she refused to cover her infant while breastfeeding.[1] As a response to this treatment, some women have begun using their breasts as a means of protest through "nurse ins," engaging in public breastfeeding en masse.

This chapter examines the rhetoric surrounding public breastfeeding and the politicization of otherwise politically uninvolved women who choose to participate in acts of politically motivated breastfeeding, or "lactivism," and considers the rhetorical dimensions of the performance of these actions. To that end, I consider two kinds of lactivism—online and in public—demonstrating the difference that a change in medium makes in perceptions of the protest action. Moreover, the case of lactivism demonstrates the difficulty in recoding the breast as maternal rather than sexual, especially as the protestors rely on the shock value of public breastfeeding to draw attention to their cause even as they claim that such an occurrence should be unremarkable. In each of these cases, commonly held stances on motherhood and femininity came to bear on the framing of the protestors, as those who breastfed in public were regarded with sympathy and support while those who chose to display their breastfeeding photos online were portrayed as self-indulgent exhibitionists.

Breastfeeding and Society

Although some have raised concerns about the amount of environmental toxins that are passed on through mother's milk to the infant[2] and others question whether it is breastfeeding itself that garners benefits to the child,[3] there is near unanimity among researchers that breastfeeding is the optimal food for infants.[4] For example, Jose Dorea argues that even for mothers who smoke, infants "are better protected by breastfeeding than by formula feeding."[5] The infant gains a decreased risk of SIDS, diabetes, Crohn's Disease, ulcerative colitis, lymphoma, and celiac disease; protection against gastroenteritis, respiratory infection, otitis media, urinary tract infection; higher IQ; reduced obesity; and the best form of nutrition.[6] The mother gains decreased risk for breast cancer, osteoporosis and hip fracture after menopause, epithelial ovarian cancer; increased bonding; and shorter time for her uterus to return to pre-pregnancy size.[7] Other scholars note that the public as a whole stands to benefit from widespread breastfeeding by reducing both infant mortality and the need for medical care, in addition to the conservation of family financial resources, especially in developing countries.[8]

Yet in spite of the many benefits of breastfeeding, there is often difficulty finding support for breastfeeding, and scholars and breastfeeding advocates have long called for legal protection and revisions in hospital policies and health care education to facilitate this practice.[9] Scholars and mothers have also noted that mothers receive mixed messages concerning the utility of breastfeeding from the moment they are discharged from the hospital and provided samples of formula, which bears the tacit stamp of approval from an authority on health.[10] Jacqueline

Wolf notes that "Even as more women in the 1970s nursed their infants, the medical community never deemed breastfeeding the standard of care. Rather, formula feeding remained the norm and nursing became the 'best' thing to do, akin to putting cotton clothing on your baby rather than polyester—a nice touch, but unnecessary."[11] This ambiguity concerning the effectiveness of formula has also spread into the popular media with aggressive formula advertisements, to which the World Health Organization responded with the International Code of Marketing Breast-milk Substitutes.[12] Judy Hopkinson notes, "Our society systematically undercuts women's efforts to breast-feed in part by downplaying the value of breast-feeding and in part by failing to adequately support women who do choose to breast-feed."[13]

In addition to the hesitant support of the medical industry, there are also cultural obstacles to breastfeeding. Women are told that breastfeeding is the most natural, beneficial way to feed their children, yet there is little structural support for this behavior short of remaining in the home. American society still largely operates with a male-centered view of the public sphere. Workplaces are not set up to accommodate nursing mothers, nor are other public spaces, unless the woman wants to share spaces that are considered dirty, such as public restrooms.[14] The architecture of the public sphere assumes that participants will be unencumbered by children. Hopkinson observes that "women are justifiably irritated when admonished to breast-feed without the cultural support necessary to enable their success."[15]

Discomfort with the nursing mother has implications for mothers, children, families, and women and society as a whole. Lactivists seek to reclaim the practice of nursing for both health and cultural reasons. Normalizing breastfeeding is a desirable goal not only from a public health standpoint, but also as a way to redefine motherhood and women in general. Yet, as we shall see, there are problems with lactivism as a way to accomplish this goal. By calling attention to that which they wish to normalize, they send mixed messages to those who oppose public breastfeeding while undermining the message that breastfeeding should be counted among other mundane activities. Moreover, it is not the efficacy of breastfeeding that is in question, but rather the cultural construction of the female breast as a sexual object that stands in the way of widespread acceptance of public breastfeeding, and in some ways lactivism reinforces this sexualization of the breast by refusing to acknowledge its erotic nature.

Public Breastfeeding as Spectacle

What makes lactivism so powerful is the bodily presence of a woman whose sexuality is ambiguous. When a woman exposes her breasts, it is interpreted in America as a sexual act. People are used to seeing breasts in movies, on billboards, on posters, in advertisements, on the beach, in nightclubs, in magazines, and at topless bars—but not in a restaurant or at the mall—at least not in person.

Perhaps it is this implied desire or the supposed seeking for desire that shocks passersby into offense. It is incomprehensible to onlookers that the woman is using her breasts precisely as they were intended. But such discomfort stems from how American society views the breast. As Hopkinson observes, "Breasts are instruments for feeding young. We think of breasts as something to sell cars with. In this country, we have a fixation on breasts."[16]

The presence of a woman who is simultaneously sexualized and maternal is shocking to many. Cindy Stearns writes, "Given the strong cultural preference for sexualized breasts, women who breastfeed are transgressing the boundaries of both the good maternal body and woman-as-(hetero)sexual object."[17] By breastfeeding in public, women are calling attention to their embodiment. When one sees breasts on a screen or a photograph, it is easy to forget the reality of the woman in the image—and the breasts connected to her. This keeps the breasts, and the woman herself, safely anonymous. When the lactivists place their embodiment on display by not only showing but *using* their breasts, they shatter the common experience of observing women's passive breasts.

Far from simply refusing to be civil by hiding their embodiment, lactivists engage in image events to call attention to it and bring it before the public eye.[18] The spectacle of the nurse-in functions as a critique of held in common notions concerning the use of the breast and the maternal body. Lactivists do not seem content to merely advocate for the right to breastfeed, but instead seek to redefine the mother's place in society. Rather than simply believing that women who bear children should not be imprisoned in the home, safely hidden away from delicate eyes, lactivists perform this belief.

Lactivists have tapped into a common theme in media coverage—extremes get covered, run of the mill protests do not. Even a one woman protest in front of Victoria's Secret can gain coverage if "a woman who could pass for a model parked her stroller, discreetly lifted her black tank top and nursed her four-month-old son."[19] Such an event is so out of place that it warrants coverage—and coverage is essential, especially in light of what Maxwell McCombs and Donald Shaw call the "agenda setting function" of mass media.[20] Without coverage, people will not have a chance to determine whether they support or oppose the goals of the protest because they simply will not think about the issue at all.

Nurse-ins draw on commonly recognized tactics of social movement protest that work, yet it is also its difference from "normal" protest actions that make these acts compelling. Lactivism uses the sexualized breast to gain media attention in order to desexualize the breast, exposing that which is generally hidden for erotic reasons to demonstrate the natural function of the breast, which is hidden beneath a layer of socially sanctioned fetishization. Lactivists challenge this fetishization, seeking to normalize breastfeeding, if not the breast itself. These lactivists perform the change they wish to encourage and such performances serve a rhetorical function. In her discussion of protest performance, Marcyrose Chvasta writes, "I know four things for sure: boundary-crossing influences policy-makers, carnival creates community, violence doesn't solve anything, and expressing felt anger mobilizes government more effectively than (gleefully ironic or otherwise) celebration."[21]

To get into the media requires transgressing boundaries, and this is one of the strengths of lactivism. Yet it is precisely this boundary transgression that others find so troubling. A continual theme that emerges from the discourse surrounding public nursing is that of propriety, tact, or civility. There is an assumption not only that these women should nurse their children safely out of sight, but also that by performing this mode of protest, they violate codes of decorum—they should somehow work within the system. An article covering a nurse-in at a water park states, "Jefferson Vowel, assistant aquatic supervisor, said he had no problem with the nursing. 'It's just when patrons complain. Our main thing is encouraging the modesty.'"[22] Noelle Steel, an observer at this action "said the protesters in the wading pool were going too far. 'It's a normal and natural thing, but not something you have to make an issue of. You can be tactful about it,' said Steel, who said she nursed her children when they were babies."[23] There is an implicit assumption that these women have a choice—they do not have to nurse where they are. Moreover, they should not "make an issue" out of nursing. The women in these protests, however, argue that they were not the ones who started making an issue out of it. As Diane Call, who participated in a nurse-in at Burger King, states, "It's not like I'm half naked swinging around saying, 'Hey, look at this. I have boobs.'"[24]

Franklyn Haiman suggests that "perhaps the best one can do is to avoid the blithe assumption that the channels of rational communication are open to any and all who wish to use them."[25] Public nursing is clearly a case where working within the system has proven ineffective. Even in states that have enacted laws that protect a nursing mother's right to nurse anywhere she is otherwise allowed to be, nursing mothers are still subject to harassment by patrons, management, security, and, in some cases, law enforcement. In other words, this has ceased to be a legal battle and has become a rhetorical battle. Through politically motivated public nursing, these women are working to redefine nursing, motherhood, and womanhood itself. On the other side of this battle are those who work to keep the breast sexualized and the tangible performance of embodied motherhood—nursing, rather than bottle feeding, for example—safely sequestered from the public sphere.

Framing Lactivism in the Media

The battle to define public nursing has been chronicled in the pages of newspapers as they reported on nurse-ins and other debates surrounding the practice of public breastfeeding. Although some media scholars have noted that social movements can be portrayed in unpredictable and sometimes damaging ways by the media,[26] Jackie Smith and colleagues found in their study of media coverage of protest actions that "the vast majority of news reports were judged as neutral in their coverage, and of those that revealed some spin either toward demonstrators or authorities, more tended to favor the demonstrators."[27] In the newspaper articles that report on public nursing, the media tended to be slightly in favor of the lactivists, although they also attempted to give voice to those who opposed public breastfeeding or the nurse-ins. In these news reports, some common themes emerged. The lactivists argue that: breastfeeding is not excretory; breastfeeding is not exposure or sexual; breastfeeding is natural and beneficial to their child; and they suffer from oppression by those who oppose public breastfeeding. Those who oppose public breastfeeding argue that: it is disgusting and offensive; women who breastfeed should be more modest, tactful, and/or discreet. In examining these stories, a frame emerged—that of the reluctant activist.

Breastfeeding Is Not Excretory

Despite being covered with a blanket, when a diner at a Burger King realized that Jody McIlmail was breastfeeding, she could not help but exclaim, "That's just disgusting!" McIlmail reported that "she went on in the same vein for nearly 15 minutes . . . She just started going off . . . There are places for that. I don't have to see that. She should be doing that in the bathroom.'"[28] The woman raised her voice so McIlmail would hear. Others take their discomfort with public breastfeeding to a more national audience; Barbara Walters is among those uncomfortable with public breastfeeding, and her comments about sitting next to a nursing mother on a plane during an episode of the "The View" sparked yet another nurse-in.[29]

Mackenzie Carpenter reported that "when BabyTalk Magazine published a cover photo of a baby nursing, most of those who wrote in—8,500—applauded the decision, but some readers called it 'pornography' and 'disgusting.'"[30] Susan Kane, *BabyTalk*'s editor in chief states, "I had one person write in and say, 'I wouldn't go to the bathroom in front of you, so why would you nurse in front of me?' They were equating breast milk with poop or pee! It was unimaginable."[31] Yet it isn't so unimaginable. In fact, the comparison between urination and lactation comes up fairly regularly. Seattle Washington's KOMO television commentator Ken Schram makes a similar comment, stating that although breastfeeding is natural, "so is urinating, but most folks don't up and pee in a glass jar in the middle of the mall."[32] Massachusetts state representative Robert Hargraves makes a similar comment, arguing that breastfeeding is "not something you do in public. The next thing they'll want to legalize is using urinals in public."[33]

Because breastfeeding is the ejection of fluid from the body, there is some logic to comparing it to excretory acts such as urination. Perhaps this is simply an extension of Western society's conception that anything that comes out of the body—sweat, semen, urine, blood, excrement—is automatically considered unclean at best and hazardous at worst. As such, the desire to confine women to places where such uncleanness is acceptable is reasonable from this stance. But milk is not the same as urine or any other bodily fluid. It is food, making the process more akin to eating, which is commonly done in public. In fact, Mary Lofton, spokeswoman for La Leche League International, states, "Ironically, exposure [in public breastfeeding] is not really the issue . . . It's the act of nursing that people are not comfortable with. We have a long history of bottle-feeding in our culture."[34] Americans have technologized the feeding of infants to the point where the natural seems unnatural.

Women are often given the choice to breastfeed in public restrooms or leave the premises, reinforcing the perception of uncleanness. Valarie Roney, who took part in an airport protest against Delta Airlines states, "It's every day in every city, people walking up to women feeding their babies, saying 'You should go to the bathroom,' or 'You should put a blanket over her.' They wouldn't want to eat in a bathroom or eat lunch with a blanket over their heads."[35] Heather Ring had a similar experience at a water park in Oregon: "a woman then came charging out of the office and said she would have to nurse in the bathroom or leave. Heather did not want to nurse her daughter in the bathroom. 'That's where people defecate or change their clothing. That's disgusting.'"[36] Another woman forcefully disavows any connection between excretion and nursing. Susan Light tried to nurse her child in a New York Public Library and "within seconds a female security guard screamed at her to 'take that outside,'" but "when Light protested, a male security manager told her that an exhibit in the room made it 'private property' and told her to go to a bathroom or leave." Light stated that "it's wrong that a mother should be forced to nurse in public bathrooms, which are filled with germs. Breast-feeding is not akin to urination or defecation."[37]

Although not everyone who opposes public breastfeeding explicitly connects breastfeeding with excretion, descriptors such as "disgusting" and "offensive" reveal the underlying subtext of uncleanness. Elizabeth Payne reports on a Canadian study that found "found 23 percent of mothers who didn't nurse said they found it unappealing or disgusting."[38] When asked about public breastfeeding, Melanie Barclay stated, "I think it's disgusting for women to breastfeed in public, especially if they don't cover it. I am a 23-year-old woman myself and I don't need to see some woman feeding her child by her breasts while I am out."[39] Kalen Johnson, a Starbucks patron states, "In a place where I'm eating or drinking, [breastfeeding is] the last thing I want to see."[40]

Yet Johnson seems to overlook the fact that in a place where he is eating or drinking, all he would see is another human being who is likewise eating and drinking. The connection between the breast as a sexual stimuli and the breast as source of infant food has been completely lost. Claude Lewis, writing for the Columbia, Missouri *Daily Tribune*, observes, "Apparently, this is a Jekyll and Hyde culture that wants bare breasts on the screen but not in the mall. We are now living in the 21st century—I hear—but far too many Americans are languishing in the distant past. You hear it phrased all sorts of ways: Breast-feeding is not appropriate, even offensive in public; it's highly personal, a private matter between mother and child. Some even call it 'disgusting.'"[41] And lest one think that it is men who are somehow jealous of the baby and his or her use of the breast, consider that many of the people who vehemently oppose public breastfeeding are women. In the *Boston Herald*, Margery Eagan asks, "Why are women just as hostile toward nursing mothers as men? . . . Whenever you bring up breast-feeding, you get as many, if not more, women whining about how mothers should hide this inappropriate and distasteful activity from view—or stay home until their babies are weaned."[42] Even Roxanne Roberts, a columnist who notes that she also breastfed—just not in public—argues against public breastfeeding, stating, "It's about the fragile balance of liberty and taste, questions of appropriateness and venue. It's about the slippery and ever-changing slope of social standards."[43]

Yet changing social standards is precisely what is at stake for lactivists. Jackie Smith and colleagues note that "political actors, including social movement activists, are not simply concerned with influencing *what* people think about; they also hope to shape *how* people think about social problems and their solutions."[44] By reinforcing the message that breastfeeding is *feeding* with the breast, lactivists reject the comparison between breastfeeding and urination or excretion. Yet they resist evoking a comparison to the bottle. This may be in part because of the belief that breast milk and formula are qualitatively different. Moreover, the overall aim of lactivism is to normalize breastfeeding, rather than simply feeding with breast milk. The act cannot be separated from the body of the woman; the breast is not the same as the bottle. There is a difference in the food itself and the means of delivery, which makes breastfeeding twice removed from the normalized, socially sanctioned mode of feeding. Thus, lactivists seek not only to redefine breast milk as food rather than excretion, but also to redefine the female breast as a medium through which they feed their children.

With those who oppose public breastfeeding defining it as excretory, perhaps it is illustrative to compare it to other bodily excretions. Although there are some who consider urine and feces erotic, this is not a commonly held position; most would define the use of urine or feces as perversion.[45] Yet in the case of urine, there is an odd duality. Two different kinds of fluid issue from the genitals. In the case of the penis, there is urine and semen, and in the case of the vulva, there is urine and vaginal secretions. For each sex, there is one that is considered dirty and one that is considered erotic, at least within certain situations—a man who ejaculates semen in a bedroom with a lover is considerably different than one who ejaculates in a crowded elevator. Although both acts would seem to be of a sexual nature, the first is acceptable and the second situation would likely be deemed disgusting, even by those who would consider the use of semen in the first situation erotic.

Despite the unmistakably sexual nature of the genitals, they obviously serve a dual purpose—sex and excretion. And public displays of sexuality are little better than excretion, perhaps held as more disgusting because of the cultural taboos we place on sexual behavior. Yet this binary does not cleanly map onto breast milk; it is neither sexual nor excretion. Rather, it is something completely different—food. Nor does the breast fit neatly within a sexual binary, as in the case of the penis and the vagina. In fact, Shirley Ardener notes that the breast and the penis "may in some ways be seen as having parallel attributes," noting that both are external and easily visible and "both produce (different) life-giving flows."[46] Because of this connection, the use of the breast may challenge masculine power in ways that the use of the vulva cannot. Both the penis and the vagina share a generative power, but the life-giving nature of the breast continues well after birth, challenging the sexual, and thus generative, nature of the breast. In short, the breast is sustaining rather than generative. Even so, the breast is defined as sexual, which makes lactation appear to those who oppose it as excretion, but because of the sexual nature of the breasts it is an excretion more akin to ejaculating in public than to urination which, although taboo, is commonly done in public (witness this after closing time at any bar in a college town) without those who oppose it adopting the moralistic overtones of those who oppose public breastfeeding. Because breast milk is generally considered non-erotic, in the mental binary of that which comes from sexual organs, it must then be considered dirty.[47]

Breastfeeding Is Not Exposure/Sexual

The sexual nature of breasts is more social construction than biological imperative and lactivists argue that breastfeeding is the sole purpose of the breast.[48] "I was feeding my baby," one woman stated. "It just seems so natural to me. This is what our breasts are for. They treated it like it was some kind of perverse act. They should be ashamed of themselves."[49] Chelsea Clark, who participated in a nurse-in at the Burlington, Vermont airport states, "It's about raising consciousness about our culture's sexualization of the breast. Breast-feeding needs to be supported wherever and whenever it happens. Babies don't know the meaning of 'wait.'"[50] In an article covering a protest against Starbucks organized by Lorig Charkoudian, who was asked by an employee to feed her baby in the restroom, Charkoudian observed, "When women breast-feed, you see less breast than you do in the average Coors Light ad . . . The breast is doing what it's designed to do."[51]

Yet lactivists recognize that the breast is defined as sexual and are quick to point out the hypocrisy of those who are offended by breastfeeding. At a protest of Victoria's Secret, which was part of a national nurse-in, Julie Wheelan stated, "'You're going to see less skin' on nursing mothers 'than what you see hanging on [the store's] walls. The real point is to see nursing not as something dirty.'"[52] Participants at a nurse-in at the Galleria in Houston, Texas "argued that current fashions show more of a women's cleavage than is exposed during breastfeeding."[53] Lewis notes that most breastfeeding mothers "are not interested in baring themselves in public. They do, however, see the hypocrisy here. They point out that they reveal less skin than is seen on many television shows and on many women at the beach. How is it that women can legally engage in topless dancing in strip clubs, dress scantily on television and prance around in bikinis on the beach—but a mother nursing her baby so naturally offends so many to the core?"[54] To answer this question, one can consider other forms of exposure that rely mainly on upholding a socially accepted double standard. For example, consider the case of *State v. Fly*, in which a man had mooned a woman by exposing his buttocks to her as she came up a staircase. After the courts found for Fly (the mooner), the North Carolina Supreme Court overturned the ruling.[55] But legal scholar Thomas Fowler observes that women had been walking around on the beaches of North Carolina with g-string bikinis that displayed their buttocks with little concern, a distinction that was upheld by the court even as they condemned the mooner. Fowler argues that "the legal analysis employed in *Fly* to achieve this result appears to be largely rationalization and artifice,"[56] suggesting that the court "appears to rewrite the indecent exposure statute and the relevant case law in novel and illogical ways simply to justify the exemption of the thong wearers that may populate our beaches."[57] In short, we as a society celebrate some forms of exposure and censure others.

Perhaps what is really on the line for lactivists is control over women's bodies and sexuality. One woman at a nursing rally asked, "Who gets to draw the line? . . . Is a quarter inch of breast showing too much? A half-inch?"[58] This is a vexing question in a culture that has difficulty defining what counts for indecent exposure even within the rarified sphere of the courts. Yet nudity is not always obscene and is, especially in popular culture, often celebrated, despite the hand wringing by those who think that such media will corrupt the youth. Advertisements for perfumes by Yves Saint Laurent and Tom Ford, for example, both push the limits of sexuality in advertising by displaying the naked female body. Yet despite these socially sanctioned forms of breast exposure, Bernice Hausman writes, "Breastfeeding in public has become an activity that women must argue into the law, on a state-by-state basis, as not obscene, thereby demonstrating that forms of female embodiment naturally following pregnancy and childbirth must be articulated as legal activities in a concerted way to make them practicable in the public sphere."[59]

Perhaps part of the concern surrounding public breastfeeding comes from an impulse to regulate not only what a particular individual sees, but what others may see as well, especially children. South Carolina State Representative Gloria Haskins, who breastfed her own children, argued against a bill exempting breastfeeding mothers from indecent exposure laws: "We have to think of those parents who . . . are walking in the mall and see someone nursing their child (who is) exposing themselves more than they would allow their child to see."[60] One example of this can be seen in a woman who was offended that another patron at a Burger King was breastfeeding under a blanket. She told her son, "Stop staring at her . . . I don't want you seeing that."[61] Yet what the boy saw was a woman completely covered by a blanket; her breasts were not exposed. Media effects scholars note that individuals perceive that others will be more affected by mediated messages than themselves; W. Phillips Davison calls this the "third person effect."[62] Such an effect also seems to spill over from the media into the physical world. Richard Perloff writes, "The 'third person' term derives from the expectation that a message will not have its greatest influence on 'me' (the grammatical first person), or 'you' (the second person), but on 'them'—the third persons. Individuals may overestimate the impact that media messages exert on others, underestimate media effects on the self, or both."[63] Those who oppose public breastfeeding are safely disgusted, but there are perverts and impressionable children out there who will think impure thoughts about the woman.

Worse yet, however, is the notion that it is the breastfeeding women who are being impure. California State Assemblyman Antonio Villaraigosa stated that "he was on a national radio talk show . . . and a male caller said breastfeeding was a way for women to get sexually aroused in public."[64] An editorial in the Albany, New York *Times-Union*, states, "Carol Murphy, entered the Most Inane Legislative Insight of the Year contest while explaining her opposition to the proposed legislation [protecting breastfeeding in New Jersey]. She told the Philadelphia Inquirer, 'A baby can be used as a weapon to embarrass people. There are women who do exactly that kind of thing.'"[65] Yet in Wisconsin, Supervisor Carousel Bayrd, who sponsored an ordinance that "prohibits anyone from interfering with a woman nursing an infant or pumping breast milk in a public place" in Dane County, Wisconsin, stated, "No one is nursing so they can have additional public nudity."[66] In a letter to the editor, Laura E. Olsen Dugan echoes this point: "No nursing mother wants to expose her breasts in public, and quite frankly, I've seen much more breast in public at swimming areas and in low-cut clothing than I do in most nursing mothers."[67]

Those opposing the lactivists express concern for how others in the community will perceive the breastfeeding mothers. Yet Kenneth Burke argues that "criminals either actual or imaginary may thus serve as scapegoats in a society that 'purifies itself' by 'moral indignation' in condemning them, though the ritualistic elements operating here are not usually recognized by the indignant. When the attacker chooses for himself the object of attack, it is usually his blood brother; the debunker is much closer to the debunked than others are."[68] Burke's formulation raises questions concerning who, exactly, those opposing the lactivists are concerned about. If Burke is correct, these protesters condemn the lactivists not only because of the potential perverseness of the breastfeeding women or the children that may be emotionally scarred, but also because of their own potential impure thoughts when confronted with such a display. By defining these women as the problem, those opposing the lactivists are able to convince themselves of their own righteousness. As Richard Gregg notes, "By painting the enemy in dark hued imagery of vice, corruption, evil, and weakness, one may more easily convince himself of his own superior virtue and thereby gain a symbolic victory of ego-enhancement."[69]

Those who support public breastfeeding are careful to maintain the binary between sexual and functional. Connecticut assemblywoman Susan V. John, who sponsored a bill protecting the right of mothers to breastfeed in public, states, "There's so many areas where we don't see women standing up for themselves, and it's refreshing when we do seem them. It is a natural act. It is a nurturing act. It's certainly not a sexual act."[70] Yet for some, breastfeeding can be a sexual act. One woman states that "when my baby sucked my breast I actually had orgasms."[71] Another woman notes that nursing "was even erotic! I imagine that's one of the reasons women are uptight about nursing, because it's overtly sexual and a baby doesn't have inhibitions like the rest of us."[72] Yet another woman vehemently opposes this, stating, "There is nothing at all sexual about nursing a baby! That is one of the ugliest myths, one of the biggest lies, one of the things that makes me want to *throw up* it makes me so angry."[73] The conflation of sexual and natural is often evoked by those who oppose public nursing. "I know that's a natural thing," argued South Carolina State Representative Gloria Haskins, "but we have other things that are very natural that we would not allow to be done in public."[74]

Much like the duality of the breast, breastfeeding does not have to be an either/or proposition. Some feminist scholars have pointed to the intersection between motherhood and sexuality. Iris Marion Young suggests that "A more radical move would be to shatter the border between motherhood and sexuality. . . . pointing to and celebrating breast-feeding as a sexual interaction for both the mother and the infant."[75] By denying any sensuality in breastfeeding and focusing only on the benefits to the child, lactivists maintain this dichotomy between sensuality (if not sexuality) and feeding.

In a culture that celebrates the sexualized female breast, putting forth a utilitarian view that seeks to downplay the erotic view of breasts is a difficult proposition to sell. Young argues that "to be understood as sexual, the feeding function of the breasts must be suppressed, and when the breasts are nursing they are desexualized."[76] But our view of breasts is not so cleanly defined; Young observes, "Breasts are a scandal because they shatter the border between motherhood and sexuality."[77] It is this ambiguity of sex and motherhood—the physical manifestation of a sexual being who is desexualized before our eyes—that makes breasts so troubling. Thus, it seems that the lactivists and those who oppose them will always speak past one another because each subscribes to such completely differing views of the breast that common ground may be impossible to reach.

Breastfeeding Is Natural and Beneficial

Closely related to the idea that breastfeeding is not a sexual act is the argument that lactivists are simply using the breast for its intended purpose. Anna Mauser-Martinez, who organized a nurse-in at a Westlake, Ohio Victoria's Secret store states, "It's kind of ironic that Victoria's Secret, which plasters breasts everywhere, is offended at seeing breasts used for their intended purpose."[78] Tristyn Underwood, who participated in a nurse-in at Burger King, echoes this sentiment: "It's normal. It's proper. It's what we were meant to do. People need to realize breasts are for more than selling beer."[79] Asserting the functional utility of the breast is often juxtaposed with a downplaying of the sexual nature of breasts. An editorial in the Albany, New York *Times Union* suggests that "in a society that makes as much money as ours sexualizing women's breasts, it wouldn't hurt to be reminded once in a while that breasts have other functions besides titillation."[80]

In addition to the notion that breasts are meant for breastfeeding, another theme that emerges is that breastfeeding is beneficial to the child. Karen Sandock, one of the organizers of Midwest Breast Fest observes, "People have the preconceived idea that it's easier to formula-feed than to breast-feed, and sometimes the social barriers are overwhelming."[81] Tracy Weber, one of the attendees echoes this sentiment: "It really makes me sad the number of babies that are formula-fed . . . Our culture doesn't support breast-feeding, and that's changing with events like this, but there's a lot of pressure on moms."[82] Indiana State Representative David Orentlicher, who sponsored a bill protecting a woman's right to breastfeed in public, stated, "We need to remove every possible social barrier so babies can get the nutrition they need and women can go about their lives."[83]

Themes of Oppression

A nursing mother was told that she could not breastfeed at Juice & Java in Albuquerque, New Mexico, sparking a nurse-in. In an article covering the protest, the *Albuquerque Journal* quotes from a letter sent by Sammie Stowe, the owner of Juice & Java to the mother who was told she couldn't breast-feed:

> In the letter, Stowe said she was unaware of the state law and was wrong to ask the woman to leave. However, she said health codes don't allow people to enter her shop without shoes or shirts and, "I do not believe that because a child is involved, a woman should be able to pull up her shirt and expose herself in public, covered by a blanket or not."
>
> She also said, "I was just considering the majority of my customers and loss of business because of your actions."
>
> She continued, "Some people believe the act of sex is a 'natural action' but (I) would not allow this to happen in my place of business either."[84]

Yet in the wake of the protest Carolyn Lawson, manager of the Juice & Java, seemed to backpedal, stating that "the business welcomes breast-feeding mothers" and that "if anything was said that breast-feeding mothers took offense to, it was purely a misunderstanding. The owner has no ill feelings toward anyone."[85]

Margery Eagan notes that "these stories always end the same way: a grand apology and a protest 'nurse-in' or a 'lactivist suckle in.' Somebody from La Leche League extols the benefits of nursing and reminds us that nursing is why

God made bosoms. But no one apparently learns a thing since we're quickly off to the next public nursing dust-up."[86] Perhaps the most powerful strategy that organizations use to brush aside criticism of anti-breastfeeding actions is the assertion that it was all a misunderstanding. In response to a nurse-in at Victoria's Secret, Anthony Hebron, a company spokesman, stated, "As far as we're considered, this is a misunderstanding, and we're sorry it's gotten to this. We've always been very accommodating to our customers and will continue to do that."[87] In response to a nurse-in at the Galleria Mall in Houston, "Galleria spokeswoman Connie Hascher said the incident was a misunderstanding. She said the officer only approached the nursing mother after she saw four men watching the woman as she breast-fed. She said the officer's suggestion was in consideration for Doyle-Madrid's privacy."[88] Mary Lofton, spokeswoman for La Leche League International, states, "Virtually all of the establishments relent or say, 'We didn't realize it was a law,' and send their apologies and say they will train their staff to be better informed."[89]

An article in the *New York Times* reports, "It is not illegal in any state to breast-feed in public, even without a special law, said Elizabeth Baldwin, a lawyer in Miami who helped draft legislation in Florida and other states. But some people—mothers, security guards, restaurant owners—are unclear on whether breast-feeding is allowed in public. 'There must be a course in security-guard school about harassing breast-feeding mothers,' Ms. Baldwin said wryly."[90] This constant reinforcement of the hassles that come from breastfeeding in public helps to justify the actions of lactivists and provide them with reasons to engage in disruptive behavior that transgresses social norms.

Newspaper accounts of lactivism are often accented with compelling narratives about the hardships of being a nursing mother. For example, Barbara Correa writes, "Brooke Ryan, the Kentucky mother of eight, produced a copy of a state law saying breast-feeding mothers can nurse in any location where they are allowed to be, blanket or not. But the restaurant manager said other diners were complaining, and Ryan left in tears. While Ryan's story is probably familiar to mothers who dare to nurse in a mall, park or other public place, it has hit a nerve with fed-up moms."[91] Later in the article, Correa reports,

> An Applebee's spokeswoman said the fallout from Ryan's experience has provided an opportunity for the company to make nursing mothers feel welcome. Moms say if that means having to cover up with a blanket, then it isn't within the law. "They just don't get it," Ryan said in an e-mail post about her experience. "It's like saying Rosa Parks has to be in the back of the bus and we will give her a pillow so she will be comfortable back there."[92]

Narratives such as this that draw on such revered civil rights icons as Rosa Parks provide moral justification to these actions and place lactivism within a long history of progressive causes.

Richard Gregg notes that oppression can serve a rhetorical function in that it validates the protestors' view of the world.

> In a perverse kind of way, protest rhetoric and protest behavior can force the kind of counter-reaction which is ego-gratifying to some individuals. The exhilaration obtained from calling the "power structure": "fascists," "repressive," and "violent" is one thing. To experience the cracking of heads, to see the blood, to face the wall of bayonets, or to receive some milder kind of "bust" from the "establishment" is proof for some that their views of reality, their perspectives which focus toward the establishment of self-hood are correct. The victory so obtained is symbolic, but nonetheless psychologically valid and important.[93]

Thus, oppression can be worn as a badge of honor that helps establish their dedication to the cause, which in this case is the most effective means of feeding their children. As such, they are fulfilling the pinnacle of maternal duty, which is making sure their young are taken care of even in the face of personal adversity.

These stories are then told and retold as a way to demonstrate the necessity for change in social mores. Narrative can be a powerful means of constituting a collective identity and indeed, Gregg notes that ego-function rhetoric is constitutive, helping to define one's definition of self.[94] Rhetorical scholars have noted that groups constitute collective identity through rhetoric.[95] They must do so because, as Michael McGee explains, there is no such thing as an empirically observable "people"; rather, "'the people' *are* the social and political myths they accept."[96] Walter Fisher suggests that "recounting and accounting for are stories we tell ourselves and each other to establish a meaningful life world."[97]

These narratives of oppression have rhetorical force in creating an understanding of what it means to be a mother, specifically a breastfeeding mother. L. A. Kauffman states that "identity politics express the principle that identity—be it individual or collective—should be central to both the vision and practice of radical politics. . . . Identity politics also express the belief that identity itself—its elaboration, expression, or affirmation—is and should be a fundamental focus of political work."[98] These women put forth a prescription not only for mothers, but also for how motherhood should intersect with society. For example, Cindy Anderson, who was asked at a Burger King play area to stop breastfeeding her child or to use the bathroom, organized a nurse-in in response. "This should be a child-friendly atmosphere. You shouldn't be expected to kind of banish yourself."[99] Such narratives display how lactivists see themselves within society, and this in turn gives rise to political actions that seek to change these perceptions. Craig Calhoun argues that "it is not just that others fail to see us for who we are sure we really are, or repress us because of who they think we are. We face problems of recognition because socially sustained discourses about who it is possible or appropriate or valuable to be inevitably shape the way we look at and constitute ourselves, with varying degrees of agonism and tension."[100]

The Reluctant Activist Frame

Media scholars have long abandoned the notion of pure objectivity in journalistic practice. Although this may be the norm that one strives for, how one reports an event will generally frame the event in such a way that it privileges one view or another. Framing can best be thought of much like a picture frame; it guides your eye to the desired focal point and obscures extraneous material. Sociologists Robert Benford and David Snow explain that "frames help to render events or occurrences meaningful and thereby function to organize experience and guide action."[101] Framing has practical consequences for those who seek to gain media attention. Thomas Nelson and Zoe Oxley argue that "issue framing is a basic component of political persuasion campaigns, but also surfaces in more objective political communication, such as news media coverage of a policy dispute. Journalists themselves may concoct their own frames, while powerful communication agents, including elected officials, interest groups, and lobbyists, create and promote frames with hopes of planting them in mass-media outlets."[102]

In the newspaper discourse surrounding lactivism, a common theme emerged, in which the lactivists were depicted as women who were compelled to participate because they had been the target of bigotry and injustice. I call this the "reluctant activist" frame. In these news stories, lactivists are not simply seeking attention by placing themselves in the public limelight. Rather, the women who become the focal point for nurse-ins are thrust into the spotlight because of the thoughtless and illegal actions of others. The woman who sparked the nurse-in at Juice & Java in Albuquerque "said she didn't want to disclose her name because she didn't want to become the 'poster mother for breast-feeding.'"[103] Another article describes a nurse-in at Victoria's Secret stemming from an incident in which Lori Rueger was told to leave the store or go to the public restroom in Old Navy: "Few of the women at the nurse-in, who ranged in age from early 20s to middle-age, have ever served as political activists before, but many said Rueger's troubles hit so close to home they felt compelled to join the demonstration."[104] Amy Harmon reports that the protest against "The View" was "inspired by similar events organized by a growing group of unlikely activists nationwide in the last year."[105]

Julie Wheelan, who staged a one-woman protest at her local Victoria's Secret as part of a national nurse-in campaign, states, "I don't know if I would be as adamant about the issue as I am, but he won't take a bottle. If I don't nurse him, I can't go anywhere."[106] An article covering a nurse-in at the Galleria in Houston observed, "Many of the mothers have stories to tell about the discomfort of breast-feeding in public. Some say they have received stares of disapproval at restaurants, museums and sporting events. Some have even had critical comments voiced loudly enough for them to hear. But nothing has made them stop. And for many, Thursday was the first time they were moved to protest."[107]

While dining at a restaurant, Adrianna Brown was told by an employee that "her breast-feeding was offensive and gave her three choices: Go to the rest room, return to her car or leave altogether. 'I flat-out refused them all,' said Brown, who was nursing her 2-month-old son at the time. 'I had every right to be there . . . and I didn't budge. But when it was all over, I felt really sad—not just for my child, but for society.'"[108] Yet the article notes that Brown "does not revel in her role as provocateur. She is simply unwilling to 'compromise' her children's health just because it offends some people's sensibilities. 'Besides, who wants to eat in a toilet?'"[109]

All of these narratives reinforce the idea that these women were driven to protest action; they did not seek out publicity, but rather, by standing up for what they believe is right, they had it thrust upon them. This helps to paint a sympathetic picture of the lactivists—they were simply minding their own business when some boorish individual harassed them, often in violation of the law. This has significant implications for the discourse surrounding public breastfeeding, as well as the way in which lactivists construct their collective identity.

Rhetorical Implications of Lactivism in Public

The presence of the exposed sexualized breast in one's proximity seems so disturbing to some individuals that even knowing that a woman is breastfeeding under a blanket can be too much to bear. This reveals not only discomfort with breastfeeding as a concept, but also a strong desire to discipline and control female bodies. The fact that women seem just as vehemently opposed to public breastfeeding as men, if not more so, demonstrates the extent to which these women have internalized the belief that the breast is solely there as a sexual object, a way to attract men (or women). As such, exposure of the breast is akin to the exposure of the genitals in public, which likewise bears a strong taboo.

But the breast *is* a sexual object. It is also a functional object, much like any other organ in the body. In denying the erotic nature of the breasts, the lactivists do a disservice to their movement. Focusing only on the function of the breast does little to address the real concerns of those who oppose public breastfeeding. Much as those who oppose public breastfeeding prescribe a way of being that hides breastfeeding in places such as restrooms and the home, with the implicit command that the breast feeders bend to the norms of society, lactivists likewise argue that people should simply get used to public breastfeeding, that the problem lies not with the nursing mothers, but with those who are offended by the practice of public nursing. As such, both are trapped in a battle of wills, each hoping that the other will simply accept their view of the situation.

By engaging in image events such as nurse-ins, lactivists have a greater opportunity to turn public opinion toward their cause. Although there are those who oppose public breastfeeding, the narratives in the newspapers betray a definite slant toward the lactivists. Indeed, it is quite easy to dismiss those who oppose public nursing as puritan cranks who hold to Victorian standards of propriety. As a father who is quite supportive of breastfeeding, I found myself sympathetic to the cause, placing myself and my wife in the place of those who occupied the newspaper narratives of oppression. Yet even those who are not sympathetic to the idea of public nursing must be aware of the proposed image of their stance by the news media, especially the editorials that paint such opposition as ignorant and reactionary. As such, newspaper narratives may help to normalize the idea of public breastfeeding as much as, if not more than, the nurse-ins themselves.

By engaging in nurse-ins, lactivists are able to shape not only public opinion about breastfeeding, but also to create a collective identity, complete with compelling narratives and a sense of oppression among women in general. The collective identity is crafted to include not only nursing mothers, but all women because breastfeeding is inextricably connected to female body experience. Lactivists seek not only to redefine breastfeeding, but to redefine the female body and what can be done with it in the public sphere. The very act of displaying the exposed breast—taboo in our culture—is not only an act of defiance, but a prescriptive act. This is an implicit argument not only for breastfeeding, but *against* bottle feeding. In seeking to normalize breastfeeding, they implicitly argue for de-normalizing formula feeding. In other words, the implicit argument is that all women should breastfeed. As such, it is an argument for all women that redefines motherhood itself.

These women, who otherwise would not be political activists, have decided to join a cause that directly affects them. In doing so, they are able to participate in acts of exposure that they would otherwise not do on their own. There is some safety in collective action, as generations of social movements have demonstrated. By aligning themselves with a group of lactivists, they are able to become only one of many women who publicly breastfeed not only out of necessity but in a show of solidarity. Breastfeeding becomes a means of building a community of motherhood rather than merely an isolated instance of exposing one's breasts.

Transgressing norms of public exposure is a powerful means of shaming establishments into acceding to their demands. Lewis writes that business owners "want the right to ask breast-feeding moms to either cover up or move on to a more private setting. But the shopkeepers do so at their own peril. They have to put up with the lactivists' ultimate weapon: the 'nurse-in,' where a group of moms breast-feed their infants in front of an establishment that directs mothers to feed their babies elsewhere. Bad publicity."[110] The use of public exposure as a means of shaming others into meeting one's demands is a practice that crosses cultures.[111]

Yet despite the potential for positive social change, one is left with a troubling question. The argumentative framework is clear—women should be free to breastfeed wherever and whenever they are called upon to do so by the hungry child; breastfeeding is a normal function that is not dirty or disgusting; the breast's functionality should trump its status as a sexual attractor. But by using the framework of spectacle, they undercut attempts to normalize breastfeeding by calling attention to the very act which they claim should be viewed as natural and mundane. They use the shock of the exposed sexual object—fully aware of the perceived sexual nature of the breast—and then suggest to the viewer that it is his or her own fault if they are shocked. Such a tactic seems counterproductive at best and likely to entrench the beliefs of those who oppose public breastfeeding.

Changing the Venue: The Case of Facebook

As noted earlier, some seem to have a visceral reaction to women who are engaging in public breastfeeding, partly because of the ambiguous nature of the exposed breast. Where the lactivists consider the function of the breast, opponents seem to consider the sexual nature of the breast. The ambiguity of nudity can make it difficult to recognize how it is functioning in a particular situation.[112] Yet I suggest that the confusion over the nature of the breast is at least partly over the space in which it is displayed. People generally seem to agree than public displays of sexual organs are improper; this belief is not merely a matter of etiquette, but is also codified into indecent exposure laws. However, as noted earlier, many of these laws explicitly exclude breastfeeding, protecting the mother's right to feed her child. Thus, if the problem is partially a matter of space in which the exposure takes place, what happens when the exposure takes place in a space where the breast is ubiquitous? In physical space, there are relatively few places one may see exposed breasts, but on the internet, they are only a mouse click away. Yet despite the relative ease of finding images of women displaying their breasts, some sites have made it against their rules to display such images. Facebook is one such site.

Facebook is an internet-based social networking site with over five hundred million users.[113] As Facebook has evolved from a college-based application to one that is community-based, there have been some clashes between those who maintain that the culture of Facebook should be that of the traditional college student and those who are older and have their own families. One area where this slippage between cultures has come to the forefront is in the controversy over breastfeeding photos on Facebook. This section considers a protest that took place both on Facebook and at Facebook's headquarters in response to the removal of photographs in which the women were breastfeeding.

The Power(?) of Online Protest

Protest action can no longer be placed neatly into traditional categories such as picketing, marching, sit-ins, and the like. With the widespread adoption of the internet, protest actions have begun to sprout up online as well. Winn Schwartau writes, "Thirty years ago, a demonstration or protest required organization and the congregation of huge numbers of people, all within the limits of the necessary police permit. Signs and slogans and chants prefaced the occasional Mayor Dalylike headline-grabbing overreactions. Today, the netherworld of cyberspace offers an unrestricted, unregulated and certainly unorganized refuge as an alternative to conventional assembly. Cyberspace provides the ideal mechanism for cyber-civil disobedience, the protest means of choice for the Information Age."[114] Other scholars have likewise suggested that such protest is inevitable. Stefan Wray states that in electronic civil disobedience (ECD), "The same principles of traditional civil disobedience, such as trespass and blockage, will be applied, but more and more these acts will take place in electronic or digital form: The primary site for ECD will be in cyberspace."[115] Digital protest has become an extension of traditional protest actions.

Some have suggested that cyberspace is changing the public sphere. Zizi Papacharissi states, "As public space, the internet provides yet another forum for political deliberation. As public sphere, the internet could facilitate discussion that promotes a democratic exchange of ideas and opinions."[116] Jan Fernback also argues that "cyberspace is essentially a reconceived public sphere for social, political, economic, and cultural interaction."[117] But one cannot ignore the differences between the digital public sphere and the physical public sphere because the medium through which protest actions take place will influence what can be done in that medium. Dick Morris suggests that "Every medium has an essence, an intransitive essence, that it communicates regardless of the content that is being sent out over that medium. . . . If you understand what that essence is as a new medium comes on the political scene, you can exploit it to be far ahead of your rival."[118] This seems similar to Marshall McLuhan's adage "the medium is the message," but Morris notes that understanding the nature of the medium is a key factor in using it for political ends.[119] After all, technology is never neutral. As Neil Postman argues, one may gain something and lose something in the bargain.[120] Diana Saco puts it another way: "If cyberspace, in one respect, can break down (physical) barriers, allowing us to roam freely across an electronic frontier, it can also erect new (digital) barriers, both in terms of who gains access and of what can be accessed."[121]

Some are quite positive concerning the potential for cyberprotest. John Clark and Nuno Themudo suggest that people are now participating in online social movement activity, or, as they put it, "dot causes": "The Internet has allowed previously opaque processes of intergovernmental negotiations to become demystified and accessible. Citizens now feel quite well informed and empowered to intervene in these processes. Dot causes provide the opportunity."[122] One can also connect with like-minded individuals in ways that transcend time and space. In her study of internet usage, Maria Bakardjieva found that "representatives of disenfranchised groups . . . were using the technology as a tool to carve spaces of sociability, solidarity, mutual support and situated, appropriative learning in communion with others."[123] Online actions can promote group solidarity among those who may never physically meet. Others have noted that individuals and groups "have used emergent technologies in order to help to produce new social relations and forms of political possibility."[124]

Although there is significant potential for protest actions online, some scholars have been more uncertain of its true libratory potential. David Gunkel explains that "participation in the virtually utopia of cyberspace, where there is supposedly no race, gender, or class, requires that one also negotiate the 'old boy network' that already dominates, informs and configures this space."[125] Dawn Dietrich makes a similar point, arguing that "women stand to gain little as quasi-disembodied subjects within a network environment *without reference to the material conditions of their subjectivity.*"[126] But even if one protests online, there is the possibility that the action may be buried under the sheer volume of information that already exists there. As Steve Jones observes, "Perhaps it is the case that the Internet allows us to shout more loudly, but whether our fellows listen, beyond the few individuals who may reply, or the occasional lurker, is questionable, and whether our words will make a difference is even more in doubt."[127] Even Wray, who is quite optimistic concerning online protests, concedes that "Electronic civil disobedience is not likely to be an organizing tool, and the end result of the ECD is not likely to be an increase in the ranks of the disaffected. Rather ECD appears to be a means to augment or supplement existing organizing efforts, a way to make some noise and focus attention."[128]

But still, there are considerable advantages to protesting online rather than in physical space. First, one is not exposed to physical harm or immediate arrest. This is particularly salient for women who may, because of the nature of their protest, want or need to bring their children along. Likewise, one need not even be who he or she is in physical space. Saco explains that

> the *true digital persona* . . . exploits the bodiless character of electronic space, allowing one to create one's own alternative identity: indeed, a nonidentity vis-à-vis the embodied individual who constructs it inasmuch as the digital persona need bear no resemblance to one's embodied self. Because online encounters are not face-to-face, none of the usual physical traits and the cultural meanings attached to those traits (e.g., gender, race, affluence) need come into play in our online practices unless we choose to identify ourselves in those terms.[129]

In essence, one can be whoever he or she wishes. Finally, one need not physically attend a protest at a specific place or time. In the case of breastfeeding mothers who have recently given birth, this is, perhaps, the most important advantage to online protest.

Another potential benefit of protesting online is the generation of social capital among those who otherwise would not physically be able to meet. Robert Putnam defines social capital as the "connections among individuals—social networks and the norms of reciprocity and trustworthiness that arise from them."[130] Social networks like Facebook help facilitate the creation of social capital among like-minded individuals by placing the emphasis on the connections between people. But Facebook also allows individuals to be connected to ideas as well as to other individuals. Some theorists have argued that this focus on ideas rather than individuals is the key to a vibrant public sphere. Richard Sennett argues that "the extent to which people can learn to pursue aggressively their interests in society is the extent to which they learn to act impersonally."[131] In his study of the eighteenth century bourgeois public sphere, Jürgen Habermas likewise celebrates the bracketing out of individual differences in arguments, yet notes that although this ideal public sphere was never actually realized, it was, nevertheless, the ideal.[132] As these women come together in the public sphere, they make connections based not on prior friendship, but on a shared passion for breastfeeding. Although such associations may seem tenuous, sociologist Mark Granovetter notes that there is significant strength in such "weak ties."[133] These online connections can have significant power in creating social capital. In the case of breastfeeding, this is particularly significant; one study found that women with more social capital were more likely to initiate breastfeeding.[134] Because of its potential for generating social capital, the connections forged through Facebook may help galvanize these women's support for breastfeeding.

Facebook, however, is not truly a public space and this case illustrates some potential pitfalls of attempting to make the digital realm serve as a way to revitalize the public sphere. A key element here is the nature of *publics*. It seems that some who have celebrated the internet as virtual public sphere have fallen into the trap described by Nancy Fraser in using the term public sphere "to refer to everything that is outside the domestic or familial sphere. Thus 'the public sphere' on this usage conflates at least three analytically different things: the state, the official economy of paid employment, and arenas of public discourse."[135] It is this last facet that is important to rhetorician Gerard Hauser, as he argues that public spheres are "discursive spaces where society deliberates about normative standards and even develops new frameworks for expressing and evaluating social reality."[136] Indeed, Michael Warner suggests that "a public is a space of discourse organized by nothing other than the discourse itself.[137] Yet despite the capacity for active deliberation, a private site like Facebook does not entirely fulfill Hauser's requirements. A quasi-public space like Facebook still retains notable limitations on what can be expressed in this environment and what kinds of frameworks can be proposed. Moreover, if the public is the discourse, as Warner proposes, what happens to the public sphere when it can be disintegrated with the touch of a button by removing the participants from that space? Such are the differences between the online world and the physical one. To remove people from a physical space requires force or, at the very least, persuasion. To erase someone's digital account, one need only hold administrator privileges. Thus despite the potential benefits of online protest, there are some significant disadvantages as well.

Tracing the Facebook Controversy

In 2008 Facebook began removing photos that featured women breastfeeding their children. This sparked outrage among women who had their profiles altered who went on to form groups such as "Hey Facebook, breastfeeding is not obscene!" and "If breastfeeding offends you, put a blanket over YOUR head!" Even an artist who painted pictures of nursing mothers received the following email:

> You posted an item that violated our Terms of Use, and this item has been removed. Among other things, content that is hateful, threatening, or obscene is not allowed, nor is content that attacks an individual or group. Continued misuse of Facebook's features could result in your account being disabled.[138]

Section seven of the "Safety" heading of Facebook's "Statement of Rights and Responsibilities" reads, "You will not post content that: is hateful, threatening, or pornographic; incites violence; or contains nudity or graphic or gratuitous violence."[139] What seemed at issue was the definition of nudity.

In response to the removal of their photos, the women staged a protest that would take place both online and in physical space outside of Facebook's headquarters. The organizers claimed that the protest was a "huge success":

> On December 27th, 2008 over 11500 people participated in our first ever M.I.L.C. (Mothers International Lactation Campaign) event. Participants from around the globe joined our virtual protest of Facebook's discriminatory practice of arbitrarily and randomly removing breastfeeding pictures from member profiles and albums, classifying them as obscene content. We raised our collective voices in opposition to Facebook by posting a breastfeeding image as our profile picture and changing our status line to: Hey Facebook, breastfeeding is not obscene! In addition to our virtual nurse-in, a live nurse-in was held at Facebook headquarters in Palo Alto CA. Both events caught the attention of media world wide. Many members received warnings and had photos removed during and after the event.[140]

But although the lactivists have staged other online protests, there has been scant media coverage since the first nurse-in.[141] This section will examine the first set of protests, both physical and online, and suggest that the successes claimed by the lactivists may in this case be somewhat overstated. The Facebook actions also demonstrate how the medium in which lactivism is performed influences the framing of the protesters and breastfeeding itself.

Framing in the Media

As with arguments mentioned in the previous discussion of nurse-ins, some of the objections to the images of breastfeeding women resort to third-person effect arguments.[142] For example, an editorial in the Everett, Washington *Herald* states:

> Reality check time: Yes, breastfeeding is beautiful. It's natural and nurturing. And the majority of people who view such a photo will see it that way. But it's not the majority of people that moms need to think about. It's the minority who can sexualize anything, or who can become obsessed with people they don't know simply because they've seen their photo. If there is a bared breast in any photo, in any context, some (or many) people will sexualize it. They are breasts after all, and this is America. . . . Private photos really shouldn't be available to the multitude of freaks the Facebook moms apparently can't begin to imagine.[143]

The writer betrays a kind of paternalism, describing the beauty of breastfeeding, while in the next breath suggesting that these mothers are naïve and must be protected from the perverts out there. Moreover, the writer alludes to the sexualization of the breast in America as part of the problem. Yet such an argument could also be made concerning the faux rape advertisements of Dolce and Gabbana found in such easily accessible places as *Cosmopolitan* and *Glamour*, which are found in almost any supermarket checkout line.

In stark contrast to the coverage of lactivism in physical space, much of the coverage in the Facebook protests sided with Facebook. One reason for this difference may be attributed to the perceived leisure of the breastfeeding mother. Where previous coverage centered on women who had little choice concerning when they would breastfeed—for example, on a flight or in a public location such as a secluded room in a library—coverage of the Facebook issue highlighted the fact that the mother chose to display breastfeeding images online. Daisy Goodwin writes,

> While the idea of Facebook as a moral arbiter is preposterous, you also have to wonder what makes a new mother decide to post a picture of herself feeding her baby for the delectation of her Facebook friends. Heather Farley, a 23-year-old from Utah who led the Palo Alto nurse-in and has been a member of Facebook for four years (400 friends), offers the explanation that as the mother of a nine-month-old, "it is hard to get a picture of me when I am not nursing." Perhaps, but doesn't that suggest she might also be too busy to post pictures of herself and her baby for the world to see?[144]

Victoria Coren argues that "The breast-feeding petitioners are obviously right, though. What an exasperating, stupid, misguided ban," but also argues that "the mother is a loony as well. Who in God's name wants to have a picture of herself breast-feeding on Facebook?"[145] In the end, perhaps it doesn't matter; Coren notes that protest actions on Facebook may be of little worth. "Clicking 'join this group' on a Facebook petition page is too easy to carry any weight. People do it for fun, or to pass the time, or by mistake. Large numbers don't make the issue important or newsworthy. One hundred thousand people have clicked to register their disapproval of the breast-feeding photo ban, but 300,000 have clicked 'I want my 90's Nickelodeon back.'"[146] Sharon Noguchi also reports, "It's hard to say whether either demonstration will move Facebook executives—who appeared to not be at work Saturday—to lift the site's prohibition of breasts displayed on members' profiles and albums."[147]

An editorial in the *Ottawa Citizen* also defended Facebook on pragmatic grounds and questioned the efficacy of the protest action: "Facebook has done so in keeping with its strict no-nudity policy, and some Facebook moms are not happy. The moms insist that there is nothing sexual and certainly not obscene about breastfeeding. What they need to understand, though, is that Facebook's policy is not so much an exercise in censorship as it is pragmatism."[148] The editorial concludes that "people should understand that a breastfeeding picture containing nudity isn't your standard 'nudie pic.' Businesses, however, have to be practical. Unless angry moms discover a way to hurt Facebook's bottom line, it's unlikely the company will change its policy, nor should it."[149]

Some argue that Facebook is being hypocritical in their selection of what passes for nudity. Anne Hart of the *Savannah Morning News* writes,

> Breast-feeding is not porn. Breast-feeding is not obscene. Breast-feeding moms should not be made to feel as though they are doing something negative that needs to be confined to a restroom. Most moms reveal much less skin while breast-feeding than the average jeans commercial reveals. Yet, the prevailing message that stems from such controversies as the Facebook ban is breasts should be hidden away when used for feeding a baby; however, displaying images of women's breasts to help sell merchandise is perfectly fine.[150]

The Minneapolis *Star Tribune* quotes Susan Kane, editor-in-chief of *Parenting* magazine, who takes a similar tack: "Breast-feeding is not nudity . . . These women are not putting tassels on their boobs."[151] However, in the seemingly public space of Facebook, that is not a question for the mothers to decide.

Protest Actions at Facebook Headquarters

Framing of the protest actions in physical space against Facebook's headquarters was generally neutral and uniform. Perhaps part of this stems from its small size. A representative report comes from Jessica Mintz of the *Charleston Daily Mail*: "At Facebook's headquarters in Palo Alto, Calif., 23-year-old mom Heather Farley, who was visiting from her home in Provo, Utah, led a real-world nurse-in to complement the online event. About 10 women showed up to breastfeed their babies outside the front door, drawing attention from local media if not Facebook employees, who were scarce on that Saturday after Christmas."[152] One reporter wrote:

> The real-life, pavement-pounding protest drew fewer placards than photojournalists Saturday, with only a handful turning out to sing, chant and breast-feed in front of Facebook's downtown Palo Alto, Calif., headquarters. But it had all the elements of a Palo Alto protest: a handful of peaceful pickets discreetly tucked away in a University Avenue plaza; placards reading "Hey Facebook, Breast-feeding is not Obscene"; protesters chatting up the media; and indifferent passers-by. A member of the Raging Grannies, the Midpeninsula activists who stage various theatrical protests, showed up to proclaim in song that "our breasts aren't porn."[153]

Other reports in the media were similarly brief.[154] It seems that the protest, although newsworthy, was not really spectacular enough to warrant much coverage.

Media coverage is an important resource for any protest. As William Gamson notes, "The media spotlight validates that they are important players. Conversely, a demonstration with no media coverage at all is a nonevent, unlikely to have any positive influence on either mobilizing potential challengers or influencing any target. No news is bad news."[155] But not all coverage is created equally. Bernadette Barker-Plummer argues that "representation in news—in fact a particular kind of representation, *voice* as opposed to *coverage*—confers authority on the source, because news itself has come to hold a special place as an authoritative version of reality."[156] But with few protestors, they were less likely to achieve this ideal. As such, they were at the mercy of the media and how they would choose to frame the action. Also, with so few protestors even showing up, the media was left with little choice concerning how they should frame it—it was amusing, perhaps, but not a serious protest. After all, within the traditional news values of impact, timeliness, proximity, and conflict, this protest was found wanting on several fronts.[157]

The protest was also crippled by timing. The span between Christmas and New Years Day is generally a slow time for business. Moreover, the social movements on people's minds generally focus on the needy, with charities dominating the scene. As such, there was little salience for the general public.

Protest Actions on Facebook

Most stories that reported on the Facebook protest tied it to the online protest. For example, Lois Collins of *Deseret News*, reports that "In December, the online protesters, who call themselves 'lactivists,' planned a virtual 'nurse in.' As they were hammering out the details, Farley asked if they wanted her to do an actual protest outside the Palo Alto headquarters while she was in California for the holidays. They did, she said, and from 11 a.m. to 2 p.m. last Saturday, she and a shifting group of about 25 others—including 10 nursing mothers over the course of the event—protested the policy. The company, she said, ignored the protest, just as it had ignored her earlier e-mails."[158] Already, one can see a distinction between the two protests with the one taking place at Facebook headquarters framed as the "actual" protest. Still, as Facebook's digital space was the legitimate site of their grievances, it made sense to bring the protest into that space.

In response to the deletion of breastfeeding photos, a group was formed called "Hey Facebook, Breastfeeding is Not Obscene!" Andrea Gordon of the *Toronto Star* reports, "Kelli Roman, a 22-year-old mother of two from Fallbrook, Calif., started the petition group in the summer after Facebook deleted several of the breastfeeding photos on her personal profile. 'It offended me,' she said over the phone from California. 'I can't see how anything about breastfeeding could be considered obscene, especially with the other things you see on Facebook . . . like scantily clad women and so much sexualized stuff.'"[159] As of June 23, 2011, the group had 258,726 members. In addition to the physical nurse in at Facebook's headquarters, they have also hosted online events, such as the virtual nurse in during the World Breastfeeding Week that took place on August 1-7, 2009. In this action, supporters were urged to show their support by replacing their profile photo with one depicting a mother nursing her child.[160] The *Seattle Post-Intelligencer* also reported that "Supporters on Facebook are asked to change their profile picture for one day to an image of a nursing mom or animal, and to change their status lines that day to 'Hey, Facebook, breastfeeding is not obscene!'"[161]

This is a fairly simple protest so far as such actions go, requiring very little of its supporters. Stephanie Muir, the organizer of Mothers International Lactation Campaign, describes the motivations surrounding the online protest: "The idea was to give some of our members an opportunity to be an armchair activist for a day and actually do something rather than just be a member of a group."[162] Yet this may be one reason why the protest action generated so little coverage; it was rather uninteresting. This protest action invited a kind of passive aggression that could be rendered impotent by the click of a mouse from a system administrator. According to sociologist Verta Taylor, "Movements act as pressure groups on behalf of people excluded from routine decision making and the dominant power structure and are crucial actors in bringing about social and political change."[163] But if this change is to take place, protestors must exert real pressure.

Rather than making creative use of the medium of Facebook, the protestors simply changed their profile pictures. As such, they did little to challenge the policies and practices of Facebook. Tiziana Terranova argues that "political intervention in an informational milieu . . . involves more than the production of counterinformation but also an engagement with the dynamics of information diffusion as such (opening up channels, selective targeting, making transversal connections, using informational guerrilla tactics)."[164] This was not done. Moreover, the knowledge that the pictures would be up for only one day allowed Facebook to simply wait for the issue to blow over. Much like protests that request that people refrain from buying gas on one day or the "Buy Nothing Day" championed by *Adbusters*, the targets of these protests know that one need only wait and people will return to buying gas or Christmas presents the very next day. In the case of Facebook, if one chooses to maintain the protest, they can simply be disconnected from the service.

By abiding by the conventions of Facebook rather than attempting to subvert them, they created a protest action that did little to gain support from the non-committed. When the only individuals likely to view one's profile are already his or her friends, such acts amount to little more than preaching to the converted. The power of traditional protest lies in the ability to take the message to those who had not already agreed to listen. For example, in considering the restriction of protest activities in residential neighborhoods, Franklyn Haiman asks, "The question, I think, is what price a society is willing to pay to insure that the messages of minority groups are not screened out of the consciences of those to whom they are addressed. For once the principle is invoked that listeners may be granted some immunity from messages they think they would rather not hear, or which cause them annoyance, a Pandora's box of circumstances is opened in which the right of free speech could be effectively nullified."[165] By engaging in protest by simply changing their own profile photo, they tacitly accepted the principle that they should not bother others, making the protest mainly self-directed. Even the protest action that took place in physical space failed to recognize this principle, as the employees were largely absent and would therefore have little chance to engage with the lactivists. Had the lactivists performed their protest at a time when employees would be present, they would have had potential opportunities to find employees sympathetic to their cause who could help to influence the existing acceptable use policy. The potential for any meaningful dialogue was squandered by poor planning and implementation.

Blending the Public and Private in Social Networking Sites

The Facebook protest itself seemed to be little more than a tempest in a teapot. However, it does demonstrate a growing sentiment that the online world should mirror, in some ways, the "real" one. Yet this ignores the fact that Facebook is private space; one does not have the rights of the landowner as a sharecropper. Charlene Li notes that "What people forget, when they say 'I do have a right, it's my page,' is that the page is provided to you as a service."[166] Yet this is cold comfort for those who have invested considerable time in building their online personae. Harvard Law School professor John Palfrey notes, "As we wrap more and more of our lives into a single environment on the Web, the feeling that civil liberties ought to be protected there continues to grow."[167] So what happened here? I suggest that three competing issues were at work in the protest over breastfeeding photos on Facebook. First, there is the ambivalence toward public breastfeeding (and breastfeeding in general) in Western society; second, there is the nature of the medium itself; and finally, there is the blending of public and private space in social networking sites that worked against these women.

In spite of the many benefits of breastfeeding, mothers who wish to breastfeed find themselves lost in a culture that has normalized bottle-feeding and formula. Because of this shift, when infants are seen eating from their natural source of food, people become uncomfortable. This is partially because the breast is seen as a sexual organ rather than a functional one. Moreover, individuals often hold contradictory attitudes toward breastfeeding. One study noted that although all of the respondents expressed a desire to have their future children breastfed, many also upheld societal restrictions on public breastfeeding.[168] When even those who recognize the benefits of breastfeeding still express discomfort with the practice, those who transgress societal norms by displaying their breasts are less likely to garner widespread support for their cause. Other groups, such as the "topfree" movement have attempted to re-define women's breasts to little avail.[169] The notion that the breast is sexual is deeply entrenched. Thus these women are seen more as exhibitionists than as mothers.

Next there is the issue of the medium. One goal for any social movement is to attract media coverage that draws attention to its cause.[170] These lactivists did attract some level of media attention, but it was unspectacular as an event. Online activism is difficult to depict in a brief news story and it is especially ill-suited to television news coverage, especially when the compelling element of the news story is that people posted photographs containing nudity. As such, the reporting was bland and uninspiring. Social movements generally use the media as a way to propel their cause into the public mind and induce some kind of sympathy, but media coverage framed it as a non-issue, one that would shortly pass. In fact, the lackluster presence of protesters in physical space, as well as the timing of the event, confirmed that reading. The online world is, by its nature, ephemeral and constantly changing. Thus, in order to signal that a movement is in for the long haul, there must be an accompanying physical presence. But online activism allows for the kind of armchair activism alluded to by one of the protestors. Bakardjieva writes, "Ordinary people . . . have not been able to afford to 'leave what you are doing now and go' . . . The handiness of the Internet access to these institutions—from amid everyday life—now generates the feeling that they are within 'attainable reach,' that people, as citizens, can actually perform action onto them, that they are part of the subjective everyday life-world and not a detached 'province of reality.' Thus 'easy' may actually mean 'it is now possible for me to act politically from where I stand.'"[171] Because many of these women were recent mothers, the ability to travel to California to protest at Facebook's headquarters would be out of reach for many of them, but the internet remained accessible.

Another aspect of the media coverage that stands in stark contrast to the coverage discussed previously is the duration of the protest. In both the physical nurse-ins and the Facebook nurse-in, the events were short in duration, but the problems leading to the protest were significantly different. In the physical actions, the women protested ongoing opposition to breastfeeding that went against their legally-held rights. Thus it was likely that the protests would continue. The Facebook protest was scheduled for one day and it seemed that the efforts would not be ongoing. In short, the Facebook protest was a story, but not one worth devoting much time to because the reporters had no reason to believe that they would see it again. Moreover, in the case of the physical protests, the law was clear—the protesters had obviously been wronged under the law, so it was an easy case to report. The status of the protagonists in the Facebook protests was more nebulous and thus invited a variety of frames.

Finally, as Daisy Goodwin observes, Facebook "blurs the line between public and private."[172] This is an intrinsic attribute of social media networks such as Facebook, which invites individuals to construct a persona for themselves. To these mothers, breastfeeding is a motherly act, a way to display and perform motherhood. To the casual observer, however, it may simply appear to be a woman with her breast hanging out with an infant in close proximity. As Kenneth Burke observed, "any selection of reality must, in certain circumstances, function as a *deflection* of reality."[173] Thus these women choose to ignore that their breasts display sexuality, while Facebook officials focused mainly on the exposure of the breast. Because Facebook owns the space, they can dictate appropriate terms of service. If one violates the terms, one may lose the ability to use the service. Such agreements are implicit in many other areas as well. For example, if one drives irresponsibly, he or she may lose the license to drive. This is generally accepted. Yet in the case of social networking sites, where the content is user-generated, users may feel considerable investment in their constructions of self.

This denial of self seems at the heart of the lactivism debate on Facebook. Rhetorician Charles Stewart notes that people in self-directed social movements "saw themselves as innocent, blameless victims, not for anything they had done to deserve victimage, but because of who or what they were."[174] When Facebook removed their breastfeeding photographs, these women were denied an integral part of their identity. As such, in these acts of lactivism, one can see traces of the "ego function" of protest rhetoric, in which the protestors engage in the action in order to gain psychic rehabilitation.[175] For Facebook, the chosen presentation of self violated the terms of service. As such, Facebook was within its rights to remove the images.

However, Facebook does not prohibit a particular presentation of self; they only prohibit its presentation on their site. In other words, they have nothing against breastfeeding itself; one simply cannot display it in that space. Herein lies the likely reason why the protest against Facebook gained little support—the women were not being oppressed. There are alternate ways to share such images if one desires. One can email them to friends, display them on their own websites (assuming they have one) or post them on social networking sites that have less stringent image restrictions. What was denied was a presentation of self in a specific space to which they had no *actual right*, merely a *perceived right*.

Stephanie Muir, the organizer of Mothers International Lactation Campaign, argues that "If women are allowed to nurse in public, they should be allowed to nurse on the Internet. . . . Young men in their 20s think it's really great to see a woman with cleavage selling beer, but as soon as you introduce a baby it gets that icky factor . . . Breasts are used for more than just selling cars."[176] In this statement, Muir conflates the Internet with Facebook, which is useful for her cause, but Facebook and the Internet are two very different things. Walter Fisher notes that narratives must be consistent and have narrative fidelity.[177] In this case, the outsiders, as well as Facebook officials, recognize that there is no dearth of breasts online, or any other body part for that matter. Finding breastfeeding images is as easy as searching for them using any search engine. Therefore the auditor is left wondering whether the problem is with breastfeeding (or breasts) or rather with where they are choosing to display these images. Although Facebook appears to be a public space, it is still a private one. Moreover, the removal of breastfeeding photos may seem reasonable to the average observer, further undermining acceptance of the lactivists' interpretation of events.

The question concerning breastfeeding photos on Facebook seems to hinge on whether Facebook is public space or private space. Such tensions are likely to become more common as more and more of our lives take place in the digital realm. Marshall McLuhan suggested long ago that electronic media serves as an extension of the central nervous system,[178] but Elaine Graham argues that "technologies are not so much an extension or appendage to the human body, but are incorporated, assimilated into its very structures. The contours of human bodies are redrawn: they no longer end at the skin."[179]

So who, then, are we? If we are to take Sherry Turkle's work at face value, we are both the digital and the flesh; both are reality.[180] Ollivier Dyens, however, argues that "the virtual being is real, but of a different kind of real, one that is both organic and technological. This being is a cultural animal, a nonorganic being. The cultural being is in a new stage of evolution."[181] I propose that it is not so simple to determine what is real, especially in light of the current fragmentation of identity. Rather, we can see in the Facebook protest a conflation of the real and the virtual, in that the protesters could not acknowledge that the two were distinct and that different rules of conduct applied in each case. Moreover, the photographs stand in as a kind of placeholder for different fragments of identity, yet all are a part of the whole. When one is deemed unacceptable, Facebook implicitly insulted the whole of the identity. Images, then, seem to function synechdocally as a representative of the whole identity, even as it is understood that it is only a part.

We see also a conflation of the public and private. Breastfeeding is generally seen as a private act, thus when the lactivists thrust their images of breastfeeding into seemingly public space, it was met with apathy at best and hostility at worst. This case illustrates the difference between true public space and perceived public space. In physical public space, women are able to breastfeed anywhere they are otherwise authorized to be. Perhaps public breastfeeding in physical space is viewed with more sympathy because the urgency of the child's hunger is a mediating factor, as anyone who has sat in a plane with a screaming, hungry child can attest. There is a function to public breastfeeding that is readily apparent—the child needs to be fed at that time and no amount of reasoning with the infant will stop the child's insistent crying. At the moment when the photograph is posted, however, the urgency has passed. As such, it appears that the woman is no longer driven by the child's insistent crying, but rather a desire to expose herself to anyone who cares to look at her profile. This may account for the lack of sympathy expressed by many in the media.

In short, one is left with the question of why these women have chosen to post images of themselves with their breasts exposed in the first place. Sociologist Erving Goffman states that for each role, there is a self-image, and that "a self, then, virtually awaits the individual entering a position; he [or she] need only conform to the pressures on him [or her] and he [or she] will find a *me* ready-made for him [or her]."[182] But sometimes the identity that we perform is not the identity that others see, or, more specifically, they do not see the role that you are performing as congruent with the performance of that role. It seems that what is at issue here is how the identity of motherhood is performed by mothers and how others think that mothers should behave. One expects online exposure from the college crowd that the lactivists inveigh against, but not from the matronly set. Moreover, it is a matter of how these behaviors are constructed within contested identities—motherhood and womanhood.[183] Where the lactivists see motherhood, Facebook sees only nudity. In other words, lactivists see the breast as a symbol, while Facebook sees the breast as simply a breast.

In the case of definitions, the odds are stacked against the lactivists. When one signs up for a service like Facebook, he or she agrees to abide by the service's Acceptable Use Policy.[184] If one violates that policy, he or she may be removed from the service at Facebook's discretion.[185] One does not have the right to the service in the same way that an individual has the right to be in the public park because Facebook is not a public space; it only seems that way. As such, the lactivist has little recourse other than to simply repost the images (which may lead to termination of the account), to leave the community, or to comply. In the end, it doesn't really matter which option the lactivists choose because they have little control over the space. As legal scholar Lawrence Lessig observes, "Code is an efficient means of regulation. But its perfection makes it something different. One obeys these laws as code not because one should; one obeys these laws as code because one can do nothing else. There is no choice about whether to yield to the demand for a password; one complies if one wants to enter the system. In the well implemented system, there is no civil disobedience."[186] In the virtual world, the system administrator is God and his or her will must be obeyed either by choice or by force.

The questions of identity online will only continue to become more complex. Likewise, the distinctions between the virtual and the physical will also continue to become less pronounced. By examining not only the successful integration of these domains, but also those in which the participants unwittingly transgress the rules of each, we will be able to more fully disentangle (or imbricate) them as each continue to evolve. Although it is tempting to simply write off such protests as mere consumer complaints against a service provider, Margaret Scammell explains, "It is no longer possible to cut the deck neatly between citizenship and civic duty, on one side, and consumption and self interest, on the other."[187] As sites such as Facebook become more integrated into our personal lives, they will likewise become integrated into our civic lives. As such, individuals and groups will continue to construct the narratives that constitute and define their lives, and the online world will be an important part of that construction regardless of policies.

Conclusion

Medical science has demonstrated unequivocally that breastfeeding is the best possible way to feed a child. Scholars and researchers have also noted that the practice of breastfeeding is not well supported in Western society for a number of reasons including the marketing triumph of formula manufacturers, difficulty in negotiating breastfeeding and work schedules, and societal squeamishness concerning the exposure of what is commonly held to be a sexual part of the body, which reflects strong taboos on overt (but not implied) sexuality. Yet if breastfeeding is to become normalized, it must enter the public sphere and lactivists have worked to make this a reality. As such, public breastfeeding has become a site of controversy that raises questions concerning the nature of the breast and its exposure, the mobility of women in society, and cultural norms surrounding motherhood and infant care.

Yet as lactivists seek to normalize breastfeeding through publicly staged image events both on and offline, they subtly undermine their message that breastfeeding should be considered natural and unspectacular. In fact, it is the very spectacle of publicly nursing women that makes such an event newsworthy in itself. A one woman protest at a mall is unlikely to raise much concern, but a beautiful woman who is publicly nursing her child in front of Victoria's Secret as a protest act is interesting enough to make the papers. Perhaps this says more about news selection practices than the nature of social movements and their tactics, but it is clear that the message that is being put forth by lactivists is "we're here, we're breastfeeding, get used to it." Calling attention to that which is done discretely—even by admission of the lactivists—makes breastfeeding a public spectacle.

It seems that neither the lactivists nor those who oppose them are willing to take seriously the opposing side's views concerning the breast. Here, then, we have a basic question of definition. So long as the lactivists doggedly hold to the stance that the breast is simply functional, while largely ignoring the stance that the breast is a sexual or an excretory organ, they are likely to simply preach to the converted, with social change coming slowly over time. In other words, the lactivists have embraced the nature of the breast, but not the culture. Peter Berger and Thomas Luckmann explain that "man is biologically predestined to construct and to inhabit a world with others. . . . In the dialectic between nature and the socially constructed world the human organism itself is transformed. In this same dialectic man produces reality and thereby produces himself."[188] One cannot ignore the cultural construction of the female breast. Eli Zaretsky argues that "the notion of identity involves negation or difference—something *is* something, *not* something else."[189] But when it comes to identity, people are multifaceted—it is this kind of binary, either/or thinking that confounds our understanding of the nursing mother. Rather than trying to determine whether or not she is sexual, natural, nurturing, or any other descriptor, perhaps the approach to consider is that she is all of these things and more. The breast is likewise all of the above. When lactivists are able to synthesize and reconcile these supposed binaries in a convincing way, perhaps it will be more difficult to oppose the practice of public breastfeeding.

This translation of ideas is essential if lactivists are to successfully alter the perceptions of those who oppose public breastfeeding. Stephen Frantzich argues that "requiring more of democracy than simply recording potentially poorly grounded preferences requires relatively open access to a wide variety of voices, allowing all ideas to be tested in the forge of open discussion. The 'winners' in such battles over words are not necessarily based on who won in the past or the loudness of the voices, but rather on the ability of the speakers to make their ideas compelling to a broad range of legitimate participants."[190] As it stands, the binary of sexual/functional is unlikely to be destroyed by acts of body rhetoric such as nurse-ins.

Lactivists must also recognize that the space in which they make their appeals matters. It is not enough to simply assert that the rights held in the physical world should automatically translate into rights held in the digital one. Especially when one cannot lay claim to the space as one can with public spaces. Although social networking sites can be an excellent means of finding like-minded individuals and building social capital, the case of Facebook also starkly illustrates A. J. Liebling's observation that "freedom of the press is guaranteed only to those who own one."

Spectacle is not enough for lactivists to make real progress in using the breast to normalize breastfeeding. What is needed is an approach that assuages the fear of the sexualized breast. After all, Cicero's observation that people "decide far more problems by hate, or love, or lust, or rage, or sorrow, or joy, or hope, or fear, or illusion, or some other inward emotion, than by reality, or authority, or any legal standard, or judicial precedent, or statute,"[191] seems just as relevant in the debate over public breastfeeding as in the ancient assemblies of Rome. More exposure, despite its effectiveness in fostering familiarity, does little to mitigate feelings of disgust or discomfort. Yet, when all is said and done, the fear lies not only with the act of breastfeeding, but with the female body itself in all of its leaky glory.[192] Shauna MacDonald notes that "we have removed bodily fluids from the intellectual/rational/masculine realm, exiling them to the material/emotional/feminine world where, paradoxically, we can ignore them by exerting control over them."[193] Those who seek to control the act of breastfeeding work also to control the female body as well. As such, the work of valuing women and the work of valuing breastfeeding seem inextricably intertwined and lactivism provides yet another entrée into this work.

Notes

1. See John Curran, "'Nurse-Ins' Support Breast-Feeding Mother Who Was Ordered Off Plane," *Daily Breeze* (Torrance, CA), November 22, 2006, A11; Grant Huang, "Breastfeeding Moms Stage 'Nurse-in' at BWI," *Maryland Gazette* (Glen Burnie, MD), November 23, 2006, A2; Associated Press, "Airline Acts on Protest by Nursing Mother; Punishes Attendant Who Ordered Her Off," *The Record* (Bergen County, NJ), November 18, 2006, A8.

2. Elizabeth Smith, "Is Breast Still Best? How Are Environmental Toxins Affecting Breast Milk?" *International Journal of Childbirth Education* 23, no. 2 (2008): 12-14. Smith notes that even with these concerns, breastfeeding really is the best option for infant feeding.

3. Joan B. Wolf, "Is Breast Really Best? Risk and Total Motherhood in the National Breastfeeding Awareness Campaign," *Journal of Health Politics, Policy & Law* 32, no. 4 (2007): 595-636. For a rebuttal of this argument, see Judy M. Hopkinson, "Response to 'Is Breast Really Best? Risk and Total Motherhood in the National Breastfeeding Awareness Campaign,'" *Journal of Health Politics, Policy And Law* 32, no. 4 (2007): 637-48.

4. Despite the significant benefits for both mother and child (see notes 6 and 7), there are some rare cases where breast-feeding is contraindicated. See Lawrence R. Berger, "When Should One Discourage Breast-Feeding," *Pediatrics* 67, no. 2 (1981): 300-02; R. C. Giglia and C. W. Binns, "Alcohol, Pregnancy and Breastfeeding; a Comparison of the 1995 and 2001 National Health Survey Data," *Breastfeeding Review* 16, no. 1 (2008): 17-24.

5. Jose G. Dorea, "Maternal Smoking and Infant Feeding: Breastfeeding Is Better and Safer," *Maternal & Child Health Journal* 11, no. 3 (2007): 288.

6. Jay Moreland and Jennifer Coombs, "Promoting and Supporting Breast-Feeding," *American Family Physician* 61, no. 7 (2000): 2093-100; Elizabeth Smith, "Breastfeeding for a Healthy Lifestyle: Is Our Education Working?" *International Journal of Childbirth Education* 22, no. 1 (2007): 30-33; S. Winterburn, "Breastfeeding Support Plans: An Evidenced-Based Approach," *Primary Health Care* 15, no. 4 (2005): 36-39.

7. D. Hollander, "Long-Term Breastfeeding, Especially of First Child, Lowers Breast Cancer Risk," *Family Planning Perspectives* 28, no. 5 (1996): 239; Moreland and Coombs, "Promoting and Supporting Breast-Feeding"; Smith, "Breastfeeding for a Healthy Lifestyle"; Winterburn, "Breastfeeding Support Plans."

8. See Edward C. Baer, "Promoting Breastfeeding: A National Responsibility," *Studies in Family Planning* 12, no. 4 (1981): 198-206; Hopkinson, "Response"; Smith, "Breastfeeding for a Healthy Lifestyle."

9. S. Bonar, "Putting Breastfeeding on the Agenda," *RCM Midwives* 10, no. 1 (2007): 13; "Breastfeeding-Related Maternity Practices at Hospitals and Birth Centers—United States, 2007," *MMWR. Morbidity And Mortality Weekly Report* 57, no. 23 (2008): 621-25; J. Clifford and E. McIntyre, "Who Supports Breastfeeding?" *Breastfeeding Review* 16, no. 2 (2008): 9-19; Deborah Jean Hales, "Promoting Breastfeeding: Strategies for Changing Hospital Policy," *Studies in Family Planning* 12, no. 4 (1981): 167-72;

Moreland and Coombs, "Promoting and Supporting Breast-Feeding"; Erin K. Murray, Sue Ricketts, and Jennifer Dellaport, "Hospital Practices That Increase Breastfeeding Duration: Results from a Population-Based Study," *Birth: Issues in Perinatal Care* 34, no. 3 (2007): 202-11; Jan Payne, Barbara Radcliffe, Emma Blank, Elizabeth Churchill, Nadia Hassan, Elizabeth Cox, and Helen Porteous, "Breastfeeding: The Neglected Guideline for Future Dietitian-Nutritionists?" *Nutrition & Dietetics* 64, no. 2 (2007): 93-98; Hila J. Spear, "Breast Is Best and Culture Wars," *International Journal of Childbirth Education* 22, no. 1 (2007): 34-35; Claire Westdahl and Sallie Page-Goertz, "Promotion of Breastfeeding—Beyond the Benefits," *International Journal of Childbirth Education* 21, no. 4 (2006): 8-16.

10. "Boosting Natural Feeding," *Midwifery Matters*, no. 111 (2006): 24-25; Deborah A. Frank, Stephen J. Wirtz, James R. Sorenson, and Timothy Heeren, "Commercial Discharge Packs and Breast-Feeding Counseling: Effects on Infant-Feeding Practices in a Randomized Trial," *Pediatrics* 80, no. 6 (1987): 845-54; Cynthia R. Howard, Fred M. Howard, Michael Weitzman, and Ruth Lawrence, "Antenatal Formula Advertising: Another Potential Threat to Breast-Feeding," *Pediatrics* 94, no. 1 (1994): 102-4; Kenneth D. Rosenberg, Carissa A. Eastham, Laurin J. Kasehagen, and Alfredo P. Sandoval, "Marketing Infant Formula through Hospitals: The Impact of Commercial Hospital Discharge Packs on Breastfeeding," *American Journal of Public Health* 98, no. 2 (2008): 290-95; Lauren Schiff, "Breastfeeding Makes for Better Health," *Mount Sinai Journal of Medicine* 73, no. 2 (2006): 571-72; Kathy Tapper, "Asserting Their Right to Feed Their Children; 266 Mothers Breast-Feed at the Cove," *South Bend Tribune* (South Bend, IN), August 3, 2003, 1; Jacqueline H. Wolf, "Low Breastfeeding Rates and Public Health in the United States," *American Journal of Public Health* 93, no. 12 (2003): 2000-10.

11. Wolf, "Low Breastfeeding Rates," 2007.

12. Annelies Allain and Kean Yeong Joo, "The Youngest Market: Baby Food Peddlers Undermine Breastfeeding," *Multinational Monitor* 29, no. 1 (2008): 17-22; Leah Frerichs, Julie L. Andsager, Shelly Campo, Mary Aquilino, and Carolyn Stewart Dyer, "Framing Breastfeeding and Formula-Feeding Messages in Popular U.S. Magazines," *Women & Health* 44, no. 1 (2006): 95-118; Diony Young, "Violating 'the Code': Breastfeeding, Ethics, and Choices," *Birth: Issues in Perinatal Care* 28, no. 2 (2001): 77-78.

13. Hopkinson, "Response," 646.

14. For more on the constraints working women face in breastfeeding, see Clifford and McIntyre, "Who Supports Breastfeeding?" 13-14; Judith Galtry, "Extending the 'Bright Line': Feminism, Breastfeeding, and the Workplace in the United States," *Gender and Society* 14, no. 2 (2000): 295-317; Laura Duberstein Lindberg, "Women's Decisions About Breastfeeding and Maternal Employment," *Journal of Marriage and the Family* 58, no. 1 (1996): 239-51; Penny Van Esterik and Ted Greiner, "Breastfeeding and Women's Work: Constraints and Opportunities," *Studies in Family Planning* 12, no. 4 (1981): 184-97.

15. Hopkinson, "Response," 646.

16. Quoted in Salatheia Bryant, "Moms Meet at Galleria for `Nurse-in' Protest," *Houston Chronicle* July 2, 2004, 27.

17. Cindy A. Stearns, "Breastfeeding and the Good Maternal Body," *Gender and Society* 13, no. 3 (1999): 309.

18. For more on image events, see Kevin Michael DeLuca, *Image Politics: The New Rhetoric of Environmental Activism* (Mahwah, NJ: Lawrence Erlbaum Associates, 1999).

19. Lynn Arditi, "Nursing Mom's Protest at Mall No Secret," *Providence Journal* July 2, 2006, B1.

20. See Maxwell E. McCombs and Donald L. Shaw, "The Agenda-Setting Function of Mass Media," *Public Opinion Quarterly* 36, no. 2 (1972): 176-87.

21. Marcyrose Chvasta, "Anger, Irony, and Protest: Confronting the Issue of Efficacy, Again," *Text & Performance Quarterly* 26, no. 1 (2006): 13.

22. Quoted in Cornelia M. Seigneur, "Moms Nurse in Protest at Water Park," *Oregonian* (Portland, OR), February 21, 1998, B6.

23. Seigneur, "Moms Nurse in Protest at Water Park," B6.

24. Quoted in Dennis Romboy, "Moms Stage 'Nurse-in' at Burger King," *Deseret News* (Salt Lake City, UT), November 11, 2003, A1.

25. Franklyn S. Haiman, "The Rhetoric of the Streets: Some Legal and Ethical Considerations," *Quarterly Journal of Speech* 53, no. 2 (1967): 114.

26. See Laura M. Arpan, Kaysee Baker, Youngwon Lee, Taejin Jung, Lori Lorusso, and Jason Smith, "News Coverage of Social Protests and the Effects of Photographs and Prior Attitudes," *Mass Communication & Society* 9, no. 1 (2006): 1-20; Carolyn Bronstein, "Representing the Third Wave: Mainstream Print Media Framing of a New Feminist Movement," *Journalism & Mass Communication Quarterly* 82, no. 4 (2005): 783-803; Kevin Michael DeLuca and Jennifer Peeples, "From Public Sphere to Public Screen: Democracy, Activism, and the 'Violence' of Seattle," *Critical Studies in Media Communication* 19, no. 2 (2002): 125-51; Todd Gitlin, *The Whole World Is Watching: Mass Media in the Making & Unmaking of the New Left* (Berkeley: University of California Press, 1980); Marie Mika, "Framing the Issue: Religion, Secular Ethics and the Case of Animal Rights Mobilization," *Social Forces* 85, no. 2 (2006): 915-41; Lynn Owens and L. Kendall Palmer, "Making the News: Anarchist Counter-Public Relations on the World Wide Web," *Critical Studies in Media Communication* 20, no. 4 (2003): 335-61.

27. Jackie Smith, John D. McCarthy, Clark McPhail, and Boguslaw Augustyn, "From Protest to Agenda Building: Description Bias in Media Coverage of Protest Events in Washington, D.C," *Social Forces* 79, no. 4 (2001): 1414.

28. Quoted in Bob Levey, "Breastfeeding in Public: 'Disgusting?'" *Washington Post* June 25, 2001, C10.

29. Amy Harmon, "'Lactivists' Taking Their Cause, and Their Babies, to the Streets," *New York Times* June 7, 2005, B3; Cecilia Kang, "Mothers Rally to Back Breast-Feeding Rights," *Washington Post* November 22, 2006, D1; Raja Mishra, "Nursing Mother's Protest Grows; Nursing Mother's Complaint Spurs Plans to Target Delta Today Organizers Target Delta Today," *Boston Globe* November 21, 2006, A1; Sue Reinert, "Nursing Moms Defend Practice; Protest Comment by Barbara Walters," *Patriot Ledger* (Quincy, MA), June 13, 2005, 1; Scott Stinson, "Barbara Walters Target of 'Nurse-in': 'Anti-Breastfeeding'," *National Post* (Don Mills, ON), June 7, 2005, A3.

30. Mackenzie Carpenter, "Bottle vs. Breastfeeding New Products Aim to Make Nursing Less Stressful, but Cultural Confusion Engulfs Moms No Matter Which Method Is Used," *Pittsburgh Post-Gazette* October 31, 2006, C1.

31. Quoted in Carpenter, "Bottle vs. Breastfeeding," C1.

32. Quoted in Debbie Cafazzo, "Lactation Intolerance," *News Tribune* (Tacoma, WA), September 24, 2006, E1. See also, Robert L. Jamieson Jr., "Pundit Who Slipped on Mother's Milk," *Seattle Post-Intelligencer* June 28, 2005, B1.

33. Quoted in "Breastfeeding Should Not Be Shunned," *The Sun* (Lowell, MA), June 25, 2007, 1.

34. Quoted in Jennifer K. Morita, "Moms Fight for Right to Nurse in Public; Despite Laws, Society Still Squeamish," *The Record* (Bergen County, NJ), August 18, 2006, A25.

35. Quoted in "Protesters Take Part in 'Nurse-Ins' at Airports," *Capital* (Annapolis, MD), November 21, 2006, A4.

36. Quoted in Margie Boule, "No Need for Tantrum over Nursing of Baby," *Oregonian* (Portland, OR), February 12, 1998, C1.

37. Quoted in Bridget Harrison, "Breast Unrest—Library Sorry for Tossing 'Milk Mom,'" *New York Post* October 2, 2002, 7.

38. Elizabeth Payne, "Still Squeamish About Breastfeeding," *Ottawa Citizen* August 19, 2008, A10.

39. Quoted in "Readers Divided on the Issue," *Toronto Star* January 30, 2007, E5.

40. Quoted in Rosalind S. Helderman, "Md. Moms Say No to Coverup at Starbucks; Women Push Chain for Policy Allowing Public Breast-Feeding in All U.S. Stores," *Washington Post* August 9, 2004, B3.

41. Claude Lewis, "You Can't Grin and Bare It in Public; Americans' Breast Obsession Doesn't Extend to Breast-Feeding," *Columbia Daily Tribune* (Columbia, MO), June 19, 2007, archive.columbiatribune.com/2007/jun/20070619comm011.asp.

42. Margery Eagan, "Let's Nip Nursing Protest in Bud," *Boston Herald* April 26, 2007, 6.

43. Roxanne Roberts, "Do Me a Favor, Keep a Lid on Your Double Latte," *Washington Post* August 11, 2004, C1.

44. Smith et al., "From Protest to Agenda Building," 1400.

45. The DSM-IV-TR lists coprophila (sexual attraction to feces) and urophilia (sexual attraction to urine) under the heading "Paraphilia not otherwise specified." See American Psychiatric Association, *Diagnostic and Statistical Manual of Mental Disorders: DSM-IV-TR*, 4th Text Rev. ed. (Washington, DC: American Psychiatric Association, 2000), 576.

46. Shirley Ardener, "A Note on Gender Iconography: The Vagina," in *The Cultural Construction of Sexuality*, ed. Patricia Caplan, 113-42 (London: Routledge, 1987), 135.

47. There are, of course, those who consider lactation sexual, but this is more commonly held as a fetish than as a typical expression of sexuality. See Brenda Love, "Lactaphilia," in *The Encyclopedia of Unusual Sex Practices*, ed. Brenda Love, 151-52 (London: Greenwich Editions, 1992). Perhaps one reason for this is because of its relative scarcity. It is, for the most part, only available from those who have recently been pregnant. However, some have induced lactation as a means of body modification and sexual expression. See Fiona Giles, "The Well-Tempered Breast: Fostering Fluidity in Breastly Meaning and Function," *Women's Studies* 34, no. 3/4 (2005): 301-26. However, even in medical discourse the conflation of all bodily fluids can be seen, especially in discussions of paraphilias. For example, Scorolli et al. lump together "Golden/brown showers, watersport, urophilia, scatophilia, lactaphilia, menophilia, mucophilia." See C. Scorolli, S. Ghirlanda, M. Enquist, S. Zattoni, and E. A. Jannini, "Relative Prevalence of Different Fetishes," *International Journal of Impotence Research* 19, no. 4 (2007): 435. Even so, the DSM-IV-TR does not seem to list lactaphilia as a paraphilia.

48. I recognize that there are evolutionary imperatives concerning breasts regarding such issues as signaling sexual fecundity and general overall health of the woman to potential mates, which play into the notion of breasts as sexual. Moreover, as a secondary sexual characteristic, breasts are an important signal of sexual development. But the breast is still seen mainly as a sexual organ because of our construction of it as such. Indeed, Judith Anderson observes that different people throughout history have found different body parts of interest and argues that "the assumption that men have always found breasts to be primarily attractive is almost certainly an artifact of our culture's

peculiar suppression of breastfeeding." Judith L. Anderson, "Breast, Hips, and Buttocks Revisited: Honest Fatness for Honest Fitness," *Ethology and Sociobiology* 9, no. 5 (1988): 320. For more on breasts and sexual selection, see Adrian Furnham, Viren Swami, and Krupa Shah, "Body Weight, Waist-to-Hip Ratio and Breast Size Correlates of Ratings of Attractiveness and Health," *Personality and Individual Differences* 41, no. 3 (2006): 443-54; Karl Grammer, Bernhard Fink, Anders P. Møller, and Randy Thornhill, "Darwinian Aesthetics: Sexual Selection and the Biology of Beauty," *Biological Reviews* 78, no. 3 (2003): 385-407; Anders Pape Møller, Manuel Soler, and Randy Thornhill, "Breast Asymmetry, Sexual Selection, and Human Reproductive Success," *Ethology and Sociobiology* 16, no. 3 (1995): 207-19; Nicolas Gueguen, "Women's Bust Size and Men's Courtship Solicitation," *Body Image* 4, no. 4 (2007): 386-90; Devendra Singh and Robert K. Young, "Body Weight, Waist-to-Hip Ratio, Breasts, and Hips: Role in Judgments of Female Attractiveness and Desirability for Relationships," *Ethology and Sociobiology* 16, no. 6 (1995): 483-507. .

49. Quoted in Boule, "No Need for Tantrum," C1.

50. Quoted in "Moms Stage 'Nurse-in' at Nation's Airports," *Deseret News* (Salt Lake City, UT), November 22, 2006, A6.

51. Quoted in "Angry Breast-Feeding Mother Organizes Starbucks 'Nurse-in,'" *Columbian* (Vancouver, WA), August 10, 2004, A8.

52. Quoted in Arditi, "Nursing Mom's Protest at Mall No Secret," B1.

53. Bryant, "Moms Meet at Galleria for 'Nurse-in' Protest," 27.

54. Lewis, "You Can't Grin and Bare It in Public."

55. See *State v. Fly*, 348 N.C. 556, 501 S.E.2d 565 (1998).

56. Thomas L. Fowler, "Of Moons, Thongs, Holdings and Dicta: State v. Fly and the Rule of Law," *Campbell Law Review* 22, no. 2 (2000): 254.

57. Fowler, "Of Moons," 258.

58. Quoted in Jamieson, "Pundit Who Slipped on Mother's Milk," B1.

59. Bernice L. Hausman, "Things (Not) to Do with Breasts in Public: Maternal Embodiment and the Biocultural Politics of Infant Feeding," *New Literary History* 38, no. 3 (2007): 491.

60. Quoted in John Frank, "Panel Approves Breast-Feeding Bill," *Post and Courier* (Charleston, SC), February 8, 2006, B1.

61. Quoted in Levey, "Breastfeeding in Public," C10.

62. W. Phillips Davison, "The Third-Person Effect in Communication," *Public Opinion Quarterly* 47, no. 1 (1983): 1-15.

63. Richard M. Perloff, "The Third Person Effect in Media Effects," in *Media Effects: Advances in Theory and Research*, ed. Jennings Bryant and Dolf Zillmann, 489-506 (Mahwah, NJ: Lawrence Elbaum Associates, 2002), 490.

64. Quoted in "Assembly Panel Again Rejects Public Breast-Feeding Proposal," *San Francisco Chronicle* May 11, 1995, A20.

65. "Right to Breast-Feed Needs Protection," *Times Union* (Albany, NY), December 10, 1996, A14.

66. Quoted in Matthew DeFour, "Law to Protect Nursing Women; County Proposal Bars Hindering Breast-Feeding," *Wisconsin State Journal* (Madison, WI), January 4, 2007, B1.

67. Laura E. Olsen Dugan Fitchburg, "City, State Should Join County in Support of Breast-Feeding," *Madison Capital Times* (Madison, WI), January 8, 2007, A7.

68. Kenneth Burke, *A Grammar of Motives* (New York: Prentice-Hall, 1945), 406-7.

69. Richard B. Gregg, "The Ego-Function of the Rhetoric of Protest," *Philosophy & Rhetoric* 4, no. 2 (1971): 82.

70. Quoted in "Mother's Outrage Prompts Bill on the Right to Nurse in Public," *New York Times* February 16, 1997, 47.

71. Daphna Ayalah and Isaac J. Weinstock, eds., *Breasts: Women Speak About Their Breasts and Their Lives* (New York: Summit Books, 1979), 238.

72. Ayalah and Weinstock, *Breasts*, 166.

73. Ayalah and Weinstock, *Breasts*, 166.

74. Quoted in Frank, "Panel Approves Breast-Feeding Bill," B1.

75. Iris Marion Young, *On Female Body Experience: "Throwing Like a Girl" and Other Essays* (New York: Oxford University Press, 2005), 89.

76. Young, *On Female Body Experience*, 88.

77. Young, *On Female Body Experience*, 88

78. Quoted in Christopher Evans, "Victoria's Secret Welcomes Breast-Feeding Protesters," *Plain Dealer* (Cleveland, OH), July 2, 2006, B1.

79. Quoted in Romboy, "Moms Stage 'Nurse-in' at Burger King," A1.

80. "Right to Breast-Feed Needs Protection," A14.

81. Quoted in Tapper, "Asserting Their Right to Feed Their Children," 1.

82. Quoted in Tapper, "Asserting Their Right to Feed Their Children," 1.

83. Quoted in Bonnie Miller Rubin, "Laws Aim to Protect Moms Who Breast-Feed," *Chicago Tribune* February 17, 2003, 1.

84. Quoted in Andrew Padilla, "Mothers Show for 'Nurse-in' Protest," *Albuquerque Journal* October 6, 1999, B1.

85. Quoted in Padilla, "Mothers Show for 'Nurse-in' Protest," B1.

86. Eagan, "Let's Nip Nursing Protest in Bud," 6.

87. Quoted in Holly Auer, "Mothers Stage Nurse-in to Support Breast-Feeding," *Post and Courier* (Charleston, SC), July 1, 2005, B1.

88. Bryant, "Moms Meet at Galleria for `Nurse-in' Protest," 27.

89. Quoted in Morita, "Moms Fight for Right to Nurse in Public," A25.

90. Quoted in "Mother's Outrage Prompts Bill," 47.

91. Barbara Correa, "Moms to Stage 'Nurse-in' at Applebee's Locations," *Inland Valley Daily Bulletin* (Ontario, CA), September 7, 2007.

92. Correa, "Moms to Stage 'Nurse-in' at Applebee's Locations."

93. Gregg, "The Ego-Function of the Rhetoric of Protest," 87-88.

94. Gregg, "The Ego-Function of the Rhetoric of Protest," 74.

95. Maurice Charland, "Constitutive Rhetoric: The Case of the *Peuple Quebecois*," *Quarterly Journal of Speech* 73, no. 2 (1987): 133-50; G. Thomas Goodnight, "Public Discourse," *Critical Studies in Mass Communication* 4, no. 4 (1987): 428-31; Michael C. Leff and Ebony A. Utley, "Instrumental and Constitutive Rhetoric in Martin Luther King Jr.'s 'Letter from Birmingham Jail,'" *Rhetoric & Public Affairs* 7, no. 1 (2004): 37-51; Michael C. McGee, "In Search of 'The People': A Rhetorical Alternative," *Quarterly Journal of Speech* 61, no. 3 (1975): 235-49; Helen Tate, "The Ideological Effects of a Failed Constitutive Rhetoric: The Co-Option of the Rhetoric of White Lesbian Feminism," *Women's Studies in Communication* 28, no. 1 (2005): 1-31.

96. McGee, "In Search of 'The People,'" 247.

97. Walter R. Fisher, "Narration as a Human Communication Paradigm: The Case of Public Moral Argument," *Communication Monographs* 51, no. 1 (1984): 6.

98. L. A. Kauffman, "The Anti-Politics of Identity," *Socialist Review* 20, no. 1 (1990): 67.

99. Quoted in Romboy, "Moms Stage 'Nurse-in' at Burger King," A1.

100. Craig Calhoun, "Social Theory and the Politics of Identity," in *Social Theory and the Politics of Identity*, ed. Craig J. Calhoun, 9-36 (Oxford: Blackwell, 1994), 20-21.

101. Robert D Benford and David A Snow, "Framing Processes and Social Movements: An Overview and Assessment," *Annual Review of Sociology* 26 (2000): 614.

102. Thomas E. Nelson and Zoe M. Oxley, "Issue Framing Effects on Belief Importance and Opinion," *Journal of Politics* 61, no. 4 (1999): 1041.

103. Padilla, "Mothers Show for 'Nurse-in' Protest," B1.

104. Auer, "Mothers Stage Nurse-in to Support Breast-Feeding," B1.

105. Harmon, "'Lactivists' Taking Their Cause, and Their Babies, to the Streets," B3.

106. Quoted in Arditi, "Nursing Mom's Protest at Mall No Secret," B1.

107. Bryant, "Moms Meet at Galleria for 'Nurse-in' Protest," 27.

108. Rubin, "Laws Aim to Protect Moms Who Breast-Feed," 1.

109. Rubin, "Laws Aim to Protect Moms Who Breast-Feed," 1.

110. Lewis, "You Can't Grin and Bare It in Public."

111. See Ardener, "A Note on Gender Iconography,"; Phillips Stevens, Jr., "Women's Aggressive Use of Genital Power in Africa," *Transcultural Psychiatry* 43, no. 4 (2006): 592-99; Terisa E. Turner and Leigh S. Brownhill, "Why Women Are at War with Chevron: Nigerian Subsistence Struggles against the International Oil Industry," *Journal of Asian and African Studies* 39, no. 1-2 (2004): 63-93. For a brief discussion of public breastfeeding as protest in Spain and the Czech Republic, see Philip Carr-Gomm, *A Brief History of Nakedness* (London: Reaktion, 2010), 99.

112. See Ruth Barcan, "Female Exposure and the Protesting Woman," *Cultural Studies Review* 8, no. 2 (2002): 62-82; Ruth Barcan, *Nudity: A Cultural Anatomy* (New York: Berg, 2004): 235-36.

113. Facebook, "Statistics," www.facebook.com/press/info.php?statistics.

114. Winn Schwartau, *Cybershock: Surviving Hackers, Phreakers, Identity Thieves, Internet Terrorists and Weapons of Mass Destruction* (New York: Thunder's Mouth Press, 2000), 70.

115. Stefan Wray, "On Electrical Civil Disobedience," *Peace Review* 11, no. 1 (1999): 108.

116. Zizi Papacharissi, "The Virtual Sphere: The Internet as a Public Sphere," *New Media & Society* 4, no. 1 (2002): 11.

117. Jan Fernback, "The Individual within the Collective: Virtual Ideology and the Realization of Collective Principles," in *Virtual Culture: Identity and Communication in Cybersociety*, ed. Steve Jones, 36-54 (London: Sage Publications, 1997), 37.

118. Dick Morris, "The Future of Political Campaigning: The American Example," *Journal of Public Affairs* 3, no. 1 (2003): 14.

119. Marshall McLuhan, *Understanding Media: The Extensions of Man* (Cambridge, MA: MIT Press, 1994), 7.

120. Neil Postman, *Technopoly: The Surrender of Culture to Technology* (New York: Vintage Books, 1993), 4-5.

121. Diana Saco, *Cybering Democracy: Public Space and the Internet* (Minneapolis: University of Minnesota Press, 2002), 210.

122. John Clark and Nuno Themudo, "The Age of Protest: Internet-Based 'Dot-Causes' and the 'Anti-Globalization' Movement," in *Globalizing Civic Engagement: Civil Society and Transnational Action*, ed. John Clark, 109-26 (London: Earthscan Publications, 2003), 117.

123. Maria Bakardjieva, *Internet Society: The Internet in Everyday Life* (London: Sage, 2005), 180.

124. Richard Kahn and Douglas Kellner, "New Media and Internet Activism: From the 'Battle of Seattle' to Blogging," *New Media & Society* 6, no. 1 (2004): 94.

125. David J. Gunkel, *Hacking Cyberspace* (Boulder, CO: Westview Press, 2001), 163.

126. Dawn Dietrich, "Refashioning the Techno-Erotic Woman: Gender and Textuality in the Cybercultural Matrix," in *Virtual Culture: Identity and Communication in Cybersociety*, ed. Steve Jones, 169-84 (London: Sage Publications, 1997), 178.

127. Steve Jones, "The Internet and Its Social Landscape," in *Virtual Culture: Identity and Communication in Cybersociety*, ed. Steve Jones, 7-35 (London: Sage Publications, 1997), 30.

128. Wray, "On Electrical Civil Disobedience," 110.

129. Saco, *Cybering Democracy*, 120.

130. Robert D. Putnam, *Bowling Alone: The Collapse and Revival of American Community* (New York: Simon & Schuster, 2000), 19.

131. Richard Sennett, *The Fall of Public Man* (New York: W.W. Norton, 1996), 349.

132. Jürgen Habermas, *The Structural Transformation of the Public Sphere: An Inquiry into a Category of Bourgeois Society*, trans. Thomas Burger (Cambridge, MA: MIT Press, 1989), 36.

133. Mark S. Granovetter, "The Strength of Weak Ties," *American Journal of Sociology* 78, no. 6 (1973): 1360-80. See also Mark Granovetter, "The Strength of Weak Ties: A Network Theory Revisited," *Sociological Theory* 1 (1983): 201-33.

134. Alex K. Anderson, Grace Damio, David A. Himmelgreen, Yu-Kuei Peng, Sofia Segura-Pérez, and Rafael Pérez-Escamilla, "Social Capital, Acculturation, and Breastfeeding Initiation among Puerto Rican Women in the United States," *Journal of Human Lactation* 20, no. 1 (2004): 39-45.

135. Nancy Fraser, "Rethinking the Public Sphere: A Contribution to the Critique of Actually Existing Democracy," in *Habermas and the Public Sphere*, ed. Craig J. Calhoun, 109-42 (Cambridge, MA: MIT Press, 1992), 11.

136. Gerard A. Hauser, "Features of the Public Sphere," *Critical Studies in Mass Communication* 4, no. 4 (1987): 439.

137. Michael Warner, *Publics and Counterpublics* (New York: Zone Books, 2002), 67.

138. Antonia Zerbisias, "Are These Obscene?" *Toronto Star* April 13, 2010, A3.

139. Facebook, "Statement of Rights and Responsibilities," www.facebook.com/#!/terms.php.

140. "Hey Facebook, Breastfeeding Is Not Obscene! (Official Petition to Facebook): Info," Facebook, www.facebook.com/#!/group.php?gid=2517126532&v=info.

141. Although there has been significantly less coverage since the first virtual nurse-in, there are still occasional reports on women who have had their breastfeeding photos removed. See Susie Mesure, "Facebook Stirs up Another Row with Nursing Mothers," *Independent on Sunday* (London), January 9, 2011, 12; Josh Stilts, "Breastfeeding Marlboro Mom Wants Answers from Facebook," *Bennington Banner* (Bennington, VT), February 1, 2011.

142. See Davison, "The Third-Person Effect in Communication"; Perloff, "The Third Person Effect in Media Effects."

143. "Some Private Photos Ought to Stay That Way," *Herald* (Everett, WA), January 5, 2009.

144. Daisy Goodwin, "A Bit Too in Your Facebook," *Sunday Times* (London), January 4, 2009, 4.

145. Victoria Coren, "7 Days: I Don't Normally Sign Petitions, but in This Case," *Observer* (London), January 4, 2009, 34.

146. Coren, "7 Days," 34.

147. Sharon Noguchi, "Online Protest of Facebook Ban on Breast-Feeding Photos Draws Tens of Thousands," *McClatchy-Tribune News Service* December 27, 2008.

148. "A Simple Policy," *Ottawa Citizen* January 14, 2009, A14.

149. "A Simple Policy," A14.

150. Anne Hart, "Hey Facebook, Breast-Feeding Isn't Obscene," *Savannah Morning News* January 4, 2009, E1.

151. Gail Rosenblum, "Feuding over PDL*; *Public Displays of Lactation; a Breast-Feeding Controversy Has Reignited the Battle among Mothers over What Is Proper While Nursing Their Babies," *Star Tribune* (Minneapolis, MN), January 31, 2009, E1.

152. Jessica Mintz, "Nursing Moms Mad over Facebook Policy," *Charleston Daily Mail* January 2, 2009, A8.

153. Noguchi, "Online Protest of Facebook Ban."

154. See, for example, Lois M. Collins, "Provo Woman Protests Facebook's Removal of Breast-Feeding Photo," *Deseret News* (Salt Lake City, UT), December 31, 2008, B2; Mike Harvey, "Mothers Angered as Facebook Cuts 'Indecent' Images; Decision by Social Networking Website to Ban Breastfeeding Photos Sparks Outrage," *Ottawa Citizen* December 31, 2008, A7; Mike Harvey, "Mothers Reveal Naked Anger as Facebook Bans 'Indecent' Images of Breast-Feeding," *Times* (London), December 31, 2008, 15; Lisa M. Krieger, "Nurse in' Planned to Protest Facebook's Ban on Photos of Lactating Breasts," *McClatchy-Tribune News Service* December 26, 2008; Jenny McCartney, "Why's Facebook So Het up About Breast-Feeding?," *Sunday Telegraph* (London), January 4, 2009, 23; Will Oremusk, "Nursing Moms Take Aim at Facebook over Breastfeeding Censorship," *Oakland Tribune* December 17, 2008.

155. William A. Gamson, "Constructing Social Protest," in *Social Movements and Culture*, ed. Hank Johnston and Bert Klandermans, 85-106 (Minneapolis: University of Minnesota Press, 1995), 94.

156. Bernadette Barker-Plummer, "News as a Political Resource: Media Strategies and Political Identity in the U.S. Women's Movement, 1966-1975," *Critical Studies in Mass Communication* 12, no. 3 (1995): 308.

157. Tim Harrower, *Inside Reporting: A Practical Guide to the Craft of Journalism* (Boston: McGraw-Hill, 2007), 17; James Glen Stovall, *Writing for the Mass Media*, 5th ed. (Boston: Allyn and Bacon, 2002), 113-15.

158. Collins, "Provo Woman Protests," B2.

159. Andrea Gordon, "Facebook Sours on Nursing Mom; More Than 10,000 So-Called 'Lactivists' Have Signed an Online Petition Protesting the Website's Policy," *Toronto Star* September 12, 2007, A1.

160. "M.I.L.C. Virtual Nurse-In," Facebook, www.facebook.com/event.php?eid=102988089741.

161. Sharon Kennedy Wynne, "Nursing Mothers Take on Facebook," *Seattle Post-Intelligencer* December 27, 2008, A10.

162. Quoted in Matt Hartley, "Breastfeeding Mothers Protest Photo Ban," *Globe and Mail* (Toronto), December 31, 2008, L3.

163. Verta Taylor, "Mobilizing for Change in a Social Movement Society," *Contemporary Sociology* 29, no. 1 (2000): 220.

164. Tiziana Terranova, "Communication Beyond Meaning: On the Cultural Politics of Information," *Social Text* 22, no. 3 (2004): 54.

165. Haiman, "The Rhetoric of the Streets," 106.

166. Quoted in Krieger, "'Nurse in' Planned to Protest Facebook's Ban."

167. Quoted in Mintz, "Nursing Moms Mad over Facebook Policy," A8.

168. Patricia Kelly Spurles and Jessica Babineau, "A Qualitative Study of Attitudes toward Public Breastfeeding among Young Canadian Men and Women," *Journal of Human Lactation* 27, no. 2 (2011): 131-37.

169. See Libby S. Adler, "A Short Essay on the Baring of Breasts," *Harvard Women's Law Journal* 23 (2000): 219-26; Helen Pundurs, "Public Exposure of the Female Breast: Obscene and Immoral or Free and Equal," *In the Public Interest* 14 (1994-1995): 1-38.

170. See Pamela E. Oliver and Daniel J. Myers, "How Events Enter the Public Sphere: Conflict, Location, and Sponsorship in Local Newspaper Coverage of Public Events," *American Journal of Sociology* 105, no. 1 (1999): 38-87.

171. Bakardjieva, *Internet Society*, 127.

172. Goodwin, "A Bit Too in Your Facebook," 4.

173. Burke, *A Grammar of Motives*, 59.

174. Charles J. Stewart, "Championing the Rights of Others and Challenging Evil: The Ego Function in the Rhetoric of Other-Directed Social Movements," *Southern Communication Journal* 64, no. 2 (1999): 101.

175. Gregg, "The Ego-Function of the Rhetoric of Protest," 74.

176. Quoted in Sharon Kennedy Wynne, "Nursing Moms to Protest Online," *St. Petersburg Times* December 26, 2008, B1.

177. Walter R. Fisher, *Human Communication as Narration: Toward a Philosophy of Reason, Value, and Action* (Columbia: University of South Carolina Press, 1987); Fisher, "Narration as a Human Communication Paradigm."

178. McLuhan, *Understanding Media*, 68.

179. Elaine L. Graham, *Representations of the Post/Human: Monsters, Aliens, and Others in Popular Culture* (New Brunswick, NJ: Rutgers University Press, 2002), 4.

180. See Sherry Turkle, *Life on the Screen: Identity in the Age of the Internet* (New York: Simon & Schuster, 1995).

181. Ollivier Dyens, *Metal and Flesh: The Evolution of Man: Technology Takes Over* (Cambridge, MA: MIT Press, 2001), 33.

182. Erving Goffman, *Encounters: Two Studies in the Sociology of Interaction* (Indianapolis: Bobbs-Merrill, 1961), 87-88.

183. How women view themselves as mothers and how others view what motherhood should be can often be very different. However, motherhood can be a powerful imperative to act, especially in matters directly related to the enactment of that role. Amy Hay notes that some of the most vocal agitators concerning the Love Canal toxic waste dump were mothers, specifically because their role as mothers was threatened by birth defects, miscarriages, and crib death. Some protestors even engaged in violent actions to force governmental agencies to listen to their demands. See Amy M. Hay, "Recipe for Disaster: Motherhood and Citizenship at Love Canal," *Journal of Women's History* 21, no. 1 (2009): 111-34. For more on motherhood and activism, see Nancy Fraser, *Unruly Practices: Power, Discourse, and Gender in Contemporary Social Theory* (Minneapolis: University of Minnesota Press, 1989), esp. chap. 7; Annelise Orleck, "'We Are That

Mythical Thing Called the Public': Militant Housewives During the Great Depression," *Feminist Studies* 19, no. 1 (1993): 147-72. For more on how the state can manage the performance of motherhood, see Eileen Boris and S. J. Kleinberg, "Mothers and Other Workers: (Re)Conceiving Labor, Maternalism, and the State," *Journal of Women's History* 15, no. 3 (2003): 90-117; Myra Marx Ferree, "The Rise and Fall of 'Mommy Politics': Feminism and Unification in (East) Germany," *Feminist Studies* 19, no. 1 (1993): 89-115; Dorit Geva, "Not Just Maternalism: Marriage and Fatherhood in American Welfare Policy," *Social Politics: International Studies in Gender, State and Society* 18, no. 1 (2011): 24-51.

184. This is not entirely unproblematic. Sandra Braman and Stephanie Roberts argue that Internet Service Providers' (ISP) Acceptable Use Policy and Terms of Service agreements are becoming a kind of de-facto law because governments worldwide are demanding more of ISPs, placing them into a regulatory role, yet allowing ISP regulations that may not be Constitutional. A similar situation seems to be at play in AUPs of services like Facebook. See Sandra Braman and Stephanie Roberts, "Advantage ISP: Terms of Service as Media Law," *New Media & Society* 5, no. 3 (2003): 422-48.

185. Facebook, "Statement of Rights and Responsibilities." Section 14 states: "If you violate the letter or spirit of this Statement, or otherwise create risk or possible legal exposure for us, we can stop providing all or part of Facebook to you. We will notify you by email or at the next time you attempt to access your account."

186. Lawrence Lessig, "The Zones of Cyberspace," *Stanford Law Review* 48, no. 5 (1996): 1408.

187. Margaret Scammell, "The Internet and Civic Engagement: The Age of the Citizen-Consumer," *Political Communication* 17, no. 4 (2000): 352.

188. Peter L. Berger and Thomas Luckmann, *The Social Construction of Reality: A Treatise in the Sociology of Knowledge* (New York: Anchor Books, 1966), 183.

189. Eli Zaretsky, "Identity Theory, Identity Politics: Psychoanalysis, Marxism, Post-Structuralism," in *Social Theory and the Politics of Identity*, ed. Craig J. Calhoun, 198-215 (Oxford: Blackwell, 1994), 200.

190. Stephen E. Frantzich, *Cyberage Politics 101: Mobility, Technology and Democracy* (New York: Peter Lang, 2002), 126.

191. Marcus Tullius Cicero, *De Oratore, Books I-II*, trans. E. W. Sutton and H. Rackham (Cambridge, MA: Harvard University Press, 1942), II.xli.178-9.

192. Elizabeth Grosz argues that "women's corporeality is inscribed as a mode of seepage. . . . they are represented and live themselves as seepage, liquidity." See Elizabeth A. Grosz, *Volatile Bodies: Toward a Corporeal Feminism* (Crows Nest, NSW: Allen & Unwin, 1994), 203. See also Margrit Shildrick, *Leaky Bodies and Boundaries: Feminism, Postmodernism and (Bio)Ethics* (London: Routledge, 1997); Ruth Tsoffar, "The Body as Storyteller: Karaite Women's Experience of Blood and Milk," *Journal of American Folklore* 117, no. 463 (2004): 3-21.

193. Shauna M. MacDonald, "Leaky Performances: The Transformative Potential of Menstrual Leaks," *Women's Studies in Communication* 30, no. 3 (2007): 341.

Chapter 4
Can You See Me Now, Driver? World Naked Bike Ride

The United States has enthusiastically embraced car culture. Even our entertainment has drawn on this love affair with the automobile, with films like *The Fast and the Furious* and *Gone in 60 Seconds*, sporting events that include NASCAR and monster truck rallies, and even a television channel named "Speed." People modify their vehicles with custom paint jobs, bumper stickers, and sometimes even more extensive modifications as seen in television shows like "Pimp My Ride." Some even turn their cars into rolling memorials with decals reading "In Loving Memory of" a particular loved one. In some ways, it seems that one's car is an extension of the self, an expression of identity. People take these cars to drive through windows at fast food joints, drive though banking, drive through pharmacies and even on vacation or road trips. Many have even engaged in sexual activities in the car.[1] For most Americans, life without a car is unthinkable.

With such an emphasis on the automobile, other forms of transportation may be edged out. When my wife and I moved to Mobile, Alabama, we chose an apartment that was located roughly in the middle of our respective workplaces. I was excited that it would be close enough to my office that I could ride my bike to work and, hopefully, get some exercise. When we arrived, however, we were surprised to find that not only were there no bike lanes, but in most places there were not even sidewalks. After driving to work for a week to get acclimated to the area, I realized that if I rode my bike to work it would only be a matter of time before I was hit by one of the many cars traveling in excess of fifty miles per hour. I had already been hit by cars twice in my life and I had no desire to be hit a third time. One of my colleagues who had moved from New Orleans explained that she had to buy a car when she moved to Mobile because the mass transit system she had relied on in New Orleans was inefficient and ineffective in Mobile.

There are, however, people who resist car culture and do so in a provocative manner. During the summer months in cities all over the world, bicyclists doff their clothing and ride naked through the streets to protest oil dependence and a car culture that leaves them vulnerable to injury. Although there have been other naked bike rides, the World Naked Bike Ride (WNBR) began as an international event in 2004 with twenty-eight cities in ten countries on four continents participating.[2] The ride was conceived of by Canadian Activist Conrad Schmidt, the founder of the Work Less Party of British Columbia.[3] The WNBR is a large, loosely organized network of such events taking place all over the globe in cities ranging from Mexico City to Portland, Oregon to Paris to London and claims to be the "world's largest naked protest against oil dependency and car culture in the history of humanity."[4]

The naming may be a bit misleading due to local statutes banning public nudity. As such, they have adopted a "come as bare as you dare" attitude. Many of the cyclists come dressed in costumes, some in body paint, and others come dressed in nothing at all. Many have signs on their bicycles while others choose to paint slogans on their exposed skin. Yet for all of them, the aim is to increase awareness of both oil dependency and the right of cyclists to share the road with cars. In examining the media coverage of these rides, however, it becomes clear that other aims are present as well.

This chapter examines newspaper coverage of the rides to explore how they are framed, as well as to examine how the protestors themselves describe the aims of the ride when given the opportunity to speak through the newspaper. The case of the WNBR illustrates the perils of maintaining a decentralized protest action as various riders and organizers highlight different facets of the ride. Still, participants in the WNBR remain united under the umbrella of opposition to car culture and cyclist rights, allowing for cohesion within the protest. Also, the nature of the protest itself functions as a powerful visual metaphor of cyclist vulnerability while maintaining a sense of absurdity that allows the protesters to make their case without appearing strident or self-righteous.

Coverage of the WNBR

For the most part, news outlets treat the WNBR as an amusing diversion. As such, the coverage ranges from short blubs that state the time and place of the planned protest to extended discussions of the legality of the events and some humorous commentary. For example, one paper compares it to another nude event: "If it's anything like most events organized by Tampa Area Naturists—TAN, get it?—there will be a half-dozen exhibitionists in thongs and 20 news photographers. Hey, somebody has to safeguard press freedom. Unless it's 40 degrees. Which could be enough to chill our First Amendment zeal."[5] The London *Times* wryly suggests, "Explain to your date that it's a protest against oil dependency and a celebration of individuality. She'll love you or leave you."[6]

Although such comments seem to trivialize the movement's goals, the riders themselves invite a humorous outlook of the protest.[7] It is difficult to be taken seriously when riding a bike adorned only with body paint and a feather boa, yet there is power in such a portrayal. Moira Smith notes that "laughter is just one of a variety of avenues open to the humor audience for showing their support of a humorous performance; the same message may be communicated with smiles, with verbal affirmations, and so forth."[8] By recognizing the amusement in the protest, the media and the observers provide tacit support.

Yet it is difficult to level the accusation of trivializing the protest when even the short blubs are true to the aims of the protest. Such coverage suggests that the news media view the protest not merely as a novelty, but as a serious protest action akin to sit-ins and marches. The London *Evening Standard* explains that "The aim is to 'celebrate sustainable transport and the human body.'"[9] The *Chicago Tribune* states that "The World Naked Bike Ride in Chicago protests oil dependency and promotes cycling and having a positive body image."[10] Some, however, oversimplified the demonstration. For example, the Vancouver, British Columbia *Sun* describes the WNBR as "an international event aimed at discouraging people from using their cars."[11]

There are a few possible reasons for the seriousness with which the news media treated the protests. First, and likely foremost, is the culture of objectivity among journalists. The role of the journalist is to report rather than comment on the action. However, there is another possible explanation—the longevity of the event. After eight years in cities around the world, it is difficult to dismiss this protest as not serious. Finally, there is the collective nature of the event. A woman in Ojai, California known as the "Pastie Lady" continually rode through town on her bicycle wearing only a hemp G-string and pasties.[12] Yet many dismissed "Earth Friend Jen's" one-woman advocacy for natural fibers, the healing power of water, and other causes. Despite the congruence of the messages—the Pastie Lady would likely be right at home in the WNBR—her messages were dismissed by those around her and she had several run-ins with the police. There is strength in numbers when it comes to protest and a large group grants the protest legitimacy that is difficult to achieve alone or with only a few individuals.

Cyclists' Description of the WNBR

For such a loosely organized protest action, the cyclists are surprisingly consistent in their statements. Five major themes emerge: the reinforcement of the goals of the protest; that the protest has little to do with exhibition; the need to attract attention; the encouragement of positive body image; and the symbolic nature of the protest. I will consider each of these themes in turn.

Reinforcing Protest Goals

Many organizers seize the opportunity to succinctly explain the purpose of their ride in terms of combating oil dependency and car culture. One London organizer states, "It's a protest against oil dependency and car culture and the overuse of cars for unnecessary reasons . . . There is too much pollution. It stinks in London, and we use too much fossil fuel."[13] Another participant states, "I was here for the ride last year. I don't like a lot of cars making smog and climate change."[14] Said another, "There are more ways to transport yourself than a car . . . This is about getting people to think about oil dependency."[15] The participants and organizers seem skilled in the art of the sound bite, recognizing that a long discourse on the benefits of bikes and the problems with cars would be unpublishable. It seems clear that participants have a clear vision of the goals of the protest and a consistent story.

Some even used the medium of letters to the editor to clarify their goals. If printed, these public letters allow the rhetor to control the discourse. After all, being interviewed for the paper or television news story does not mean that one's remarks will make it into the story, or that the reporter will include them unedited. One organizer writes, "Every June we ride our bikes to protest oil dependency. We ride naked to show how vulnerable cyclists are on the road."[16] Another participant explains that "surveys show that many of our residents and visitors would love to be able to leave their polluting cars at home and get the health benefits of riding a bike but feel unable to because of the dangerous conditions. The World Naked Bike Ride is a global event that draws attention to the need for positive changes for cyclists."[17]

In this last letter, one can sense some slippage concerning the goals of the event as the focus shifts from a protest against something (dependence on oil and climate change) to a protest for something (the rights of cyclists). The two seem intertwined, but here we see the inherent dangers of a grassroots protest event—participants may go off script. For example, despite the stated goals of the ride, some participants challenged whether this action was even a protest at all. Even the organizer of one ride in Florida employs a kind of semantic argument against applying for an event permit: "It's not really an event anyway. We're just going to go biking out there, and if some other people want to join us and bring 20 of their friends . . ."[18]

Redefining the protest as a leisurely ride may seem to absolve one of the need to apply for an event permit, but it is rhetorical suicide. Without a clear vision of the message(s) the riders wish to convey, such an event becomes merely flash with little substance. More importantly, one may lose significant rights under the law. Political speech is well-protected under the First Amendment; nude cycling would be viewed as indecent exposure or lewd conduct. Unless one is a nudist, there is little reason to bicycle naked through the streets. Without a more compelling rationale for the ride, these protesters may be seen as little more than a freak show on wheels. It is precisely this emphasis on the event as a protest action that allows both protestors and observers to focus on the message rather than on the spectacle alone.

But the arguments made go beyond those spoken at the protest. The protestors themselves embody the very actions that they desire to invoke in others, namely using bicycles and other self-propelled transportation rather than automobiles. Even the goals of body acceptance are demonstrated in the event as people of all shapes and sizes participate in the ride.

WNBR Is Not Exhibitionism

As with the stated goals of the protest, WNBR participants are sometimes conflicted concerning their own individual motives. Many emphatically state that they are not exhibitionists. One participant told the Austin, Texas *American-Statesman* that "It's not an exhibitionism thing, but you're testing the law, and you're also testing people's reaction to it," and another rider echoes this sentiment: "There are no sexual connections in this ride, and we are not exhibitionists out for a thrill. We are no different than anyone else, except we have experienced the exhilaration of feeling the wind on our skin and the utter sense of freedom from the constraints of textile coverings in the hot summer."[19] Even the press reiterates this aim. The London *Guardian* reports that "for those involved, there's more to biking in the buff than mere exhibitionism; it's also a protest against oil dependency and 'a symbol of the vulnerability of the cyclist in traffic.'"[20]

Yet within these statements, one can see an element of sensuality in the experience. Several participants stated that they were participating not only because they believed in the stated goals of the protest but also because they thought it would be fun. One organizer said, "It's like a fun protest . . . We want to have fewer cars and more room for bikes. We want to get people out of their cars and onto bikes."[21] Another rider predicted, "I think that it's going to be really fun riding around the streets half-naked in Tulsa, to be honest,"[22] and a teacher in Ireland said that she "just wanted to do something crazy."[23] Such statements call into question the purity of the claims that this is not about exhibition. For those observers subscribing to the taboo on public nakedness, it would be difficult for them to imagine finding pleasure in an activity that for many would be the pinnacle of embarrassment.

Yet the participants are in agreement; this is a fun activity. These are not self-flagellating individuals seeking social change through their own mortification and emotional pain. But this inducement of emotion in the protestors themselves plays an important role in protest action, helping to reinforce collective identity and generate enthusiasm for the cause.[24] Such self-generated desire to participate in protest actions is essential for organizers who have little control over whether volunteers will choose to participate.[25] Moreover, this illustrates further the necessity of defining the event as a protest action. Without a clear message, these smiling cyclists would be seen as merely seeking an excuse to expose themselves to an unsuspecting public.

The Desire for Attention

The assertion that the protest has nothing to do with exhibitionism stands seemingly in stark contrast to the desired goal of attracting attention. But one can hardly blame the WNBR for trying to get the attention of the media and the general public. Simon Cottle notes that although the interrelation between news media and social movement demonstrations has long been forged, "What is unprecedented is the extent to which protests and demonstrations today have become reflexively conditioned by their pursuit of media attention, and need to be if they are to get their message across and mobilize wider support."[26] As such, protestors have become increasingly savvy about getting their message out. As one ride promoter notes, "If you have a bike ride with 5,000 people, it won't get any press attention. If you have a bike ride with 50 people, and two of them are naked, it'll make the front page . . . Lady Godiva figured it out (hundreds) of years ago."[27]

This view is vindicated by the coverage the ride has received. The St. Louis *Post-Dispatch* offers the following lead for a story on the WNBR: "Picket lines? Yawn. Boycotts? Been there, done that. Cream pie in the face? With rising food costs, who can afford quality tarts? But baring it all while biking through St. Louis' bumpy streets—now that's a demonstration guaranteed exposure."[28] Even the observers noted that the tactic of riding nude would garner the desired attention: "It's weird . . . unusual . . . Maybe that will attract people's attention. I think it's a better way to protest like this."[29] The St. Louis *Post-Dispatch* reported that one observer had driven from Centralia, Illinois to St. Louis, Missouri, which is over sixty-five miles away, to view the spectacle.[30]

But it is not just about getting the attention of the public or of the media. The riders make a direct connection between getting the attention of the media and the attention of the drivers that they hope to influence. The London *Daily Mail* quotes one organizer who stated, "Many drivers don't see a place for cyclists on the road but I think they will certainly notice this."[31] Organizers in Cork, Ireland echo this point: "Cycling naked serves two purposes. It emphasises how vulnerable the cyclist is on the road, and it draws attention to the issue. A typical slogan painted on riders bodies is Can you see me now driver?"[32] These cyclists, then, use the attention that the ride will generate to suggest the kinds of attention that will allow them to share the road with other vehicles.

Promoting Positive Body Image

Like the desire for attention, the stated goal of promoting body acceptance seems slightly out of line with assertions that the ride has nothing to do with exhibitionism. This goal was stated with far less frequency than the other themes, most likely because it has always been a secondary focus of the protest, existing in the shadow of the quest for cyclist rights. But some still drew on this ideal. One participant stated that the ride should "remind us of our free spiritedness. All bodies are beautiful."[33]

Perhaps this is a case in which the mode of protest created new, unintended goals. In order to disrobe and ride naked through the streets, one must have at least some acceptance of his or her own body, and some assurance that one will not be turned away for not measuring up to some predetermined ideal. Thus the promotion of acceptance of all body types is a convenient addition to a protest action that requires one to go "as bare as you dare." Yet this cause alone would make little sense on a bicycle, and would be better suited to a nudist organization or some other group that explicitly worked to change perceptions of the body. As such, the core focus of the event remains steadfastly on issues of oil dependence and the safety of cyclists.

Nude Protest as Symbolic

One recurring theme that emerges in the discourse is the symbolic nature of the protest. The nudity in the protest is not merely a way to gain attention or to shock observers. Rather, the protesters make an explicit connection between the nudity they experience in the ride and the relative lack of protection that they feel on the road every day. The *New York Times* quotes an organizer who states, "What we clearly want to show is that we feel naked against the traffic . . . Every day we have to put up with the traffic, the aggression of the drivers, their speed and bad manners."[34] Others are even less conciliatory. One protestor told the *Toronto Star*, "People think it might be rude to be naked . . . It's a great deal ruder to be spewing toxic gas out in front of everybody's face. It's beyond rude. It's an actual health issue. What we're doing is good clean fun, and it stands in stark naked contrast to the rather obscene car culture that we've developed."[35]

Several riders described their own experiences of vulnerability when riding among cars. In New Zealand, "Third-time participant Kat Heeramaneck, of Rangihaeata, said she had almost been hit by a car twice when cycling on Bird Hill. 'I would like to feel safe when I bike to work.'"[36] Another participant in Tulsa, Oklahoma stated, "It's scary to drive on the roads . . . I've been hit by a car just recently, because of people not paying attention. I'm real excited about this big group of people being on the streets and seeing that bikers are on the streets, too."[37] This sentiment was powerfully reinforced by the next line of the article: "The local event was set for June, but it was postponed after the deaths of two cyclists killed by a motorist in Sand Springs."[38] In this news article, the journalist drives home the point that these accidents have real consequences beyond the typical fender-bender. The use of personal testimony can be a powerful tool for persuasion, especially as a way to persuade individuals that they personally are at risk.[39] As such, the primary impact of these narratives is likely to be among the riders themselves who would certainly wish to avoid being struck by a car. However, this also serves as a reminder to drivers who could envision themselves hitting a bicyclist. This may be particularly useful as a way to persuade drivers that they personally are a part of the problem because, as John De Wit and his colleagues suggest, narratives may invite less defensive responding than abstract statistical evidence.[40]

Another organizer makes an explicit connection to the nudity of the protest and the peaceful intentions: "Nudity is essential to show we're unarmed."[41] Another organizer made a similar connection, stating that "going naked is more peaceful than throwing bricks or starting fires."[42] The idea that the protesters are unarmed stands in contrast to those who imply that the motorists are really the ones that are armed. An organizer in Brighton, UK stated, "It's time more motorists stripped off their armor plating and moved around more gently on this earth . . . Bikes and naked bodies harm nobody."[43]

The sense of harmlessness comes not only from their nakedness, but also from the jovial atmosphere which results in part from the ridiculous nature of the protest action. Moira Smith states that "humorous and serious modes of discourse represent incompatible approaches to the world; where the serious attitude routinely avoids or explains away paradox, inconsistency, and ambiguity, humor embraces them."[44] One is rarely confronted with a brightly painted, nude convoy of bicyclists, so such a vision may appear to be absurd because it is so far outside of the norm. The only logical explanation the observer is left with is that it must be some kind of joke. But joking can be a serious matter and the absurd nature of the protest has significant rhetorical force. When one acts in such a way, he or she does more than simply amuse observers; the protestor sends an implicit message that he or she means no harm, and that the protest is done with a spirit of fun. In his study of anti-globalization protests, Jeffrey Juris argues that the "nonviolent performances symbolically contrasted the vulnerable, morally righteous bodies of the protestors with the menacing bodies of the police."[45] In a similar way, the stark contrast between the nude bicyclists and the armed police officers not only reinforce the protestor's vulnerability, but also set it up in the eyes of the observer as an unfair fight.

Opponents' Description of the WNBR

Opponents to the ride challenge not the motives of the protest, but rather the nudity involved. The most common argument is that such protests should be curtailed because they are harmful to any children that may be present. The

London *Daily Mirror* reports that "World Naked Bike Ride Day has left opponents fearing the Capital will now become a magnet for perverts. And [Edinburgh] Tory city councillor Alastair Paisley said: 'It is totally obscene. I would not want to see young people exposed to this sort of thing.'"[46] In another case, the City of Tampa, Florida sent a letter to the group stating, "While we protect your rights under the First Amendment . . . we believe the citizens have a right to drive their children on public streets without encountering naked adults."[47] John Palm, the organizer of the ride, told the *Tampa Tribune*, "These are the offending body parts that would end the world if they were exposed to the public."[48]

The response of the Tampa authorities illustrates the thorny nature of political nudity in the United States. The courts have long grappled with the balance between protected expression and the right of others to not have their sensibilities offended.[49] Political speech is afforded the highest level of First Amendment protection, and despite its carnivalesque nature, there can be little doubt that the WNBR is a political demonstration. Legal scholar Danielle Moriber notes that "the more nudity is intertwined or associated with the message, the more likely the speech will be protected. Nudity for nudity's sake, which is not related to the speech at issue, and use of nudity to get additional attention, are unlikely to be upheld."[50] In the case of the WNBR, the nudity functions both as a means of garnering attention and as a powerful metaphor of vulnerability. As such, it should be protected regardless of sensibilities, but whether such expression would hold up in court is uncertain.

Other countries have also struggled over what to make of the protest, and have likewise expressed concern for the children that may be playing nearby. The London *Daily Mail* quotes Susie Agnew, secretary of Marchmont and Sciennes Community Council: "It will do their cause more harm than good. There are children out playing on a Saturday afternoon and it will attract the wrong sort of crowd."[51] But exactly what sort of wrong crowd will be attracted is, of course, left for the reader to decide. As with the legislation in the United States surrounding nude dancing establishments, government officials are relegated to arguing the secondary effects of the nudity rather than the nudity itself.[52] Yet rather than explain the kind of "wrong crowd" that such protest would attract, Agnew leaves it to the readers to construct the bogeyman of their choice. Such an argument functions enthymematically, with the audience supplying the missing information in the argument. This a powerful means of persuasion; Lloyd Bitzer suggests that "since rhetorical arguments, or enthymemes, are formed out of premises supplied by the audience, they have the virtue of being self persuasive. Owing to the skill of the speaker, *the audience itself helps construct the proofs by which it is persuaded.*"[53]

What is missing in these arguments is what, exactly, the harm to children would be. After all, the protesters are merely riding bicycles; they are not engaging in anything remotely resembling sexual activity. The cyclists would be unlikely to approach any children playing nearby, as they would be too busy actually riding along the predetermined route. More fundamentally, no one questions what makes it harmful for children to observe, but harmless to adults. The people making such arguments seem to take as a given that when children observe nudity it will be harmful to them in some way. John Ince suggests that this stance may be based more on the opinions and fears of the individual making the claim than on empirical evidence: "The idea that the sight of genitals harms children is caused in part by rationalizing the genital aversion. A mother who is averse to the sight of her own genitals can make sense of that negative emotion by inventing the idea that nudity causes harm to her kids."[54] However, this matters little if a substantial amount of the audience likewise subscribes to the belief that nudity harms children.

The Rhetorical Power of the WNBR

Rhetorical scholars and sociologists alike have noted that social movements are actually about constructing (and reconstructing) meanings within a society. In fact, rhetorician Michael Calvin McGee bluntly states that a "social movement is a set of meanings and not a phenomenon."[55] Sociologist Arthur Frank likewise argues that "any sociological study has at least an implicit concern with reality construction, considered as the generation and maintenance of some organization of affairs, whether a family or a factory, a friendship or an illness."[56] It is this individual action in the service of the larger collective that makes social movements interesting phenomena. In the case of the WNBR, it is clear that most of these riders would not ride their bikes naked through the streets on their own; it is the collective nature of the action that provides license to do so. Alberto Melucci argues, "A discipline that sets out to study social movements can accomplish its task meaningfully only if it starts out from a theory that can account for the specificity and autonomy of social *action*, and can give a foundation to its *collective* character as something different from the sum total of aggregate individual behaviors."[57] From this perspective, the WNBR is not merely a menagerie of many naked bicyclists, but a mass that has its own character apart from the individuals themselves.

Alain Touraine explains, "Far from being antithetical, social movements and democracy are, in contrast, indissociable."[58] Social movements are democracy at work. This principle is so important that the rights to peaceably assemble and to petition the government for a redress of grievances are included within the First Amendment of the Constitution. But to mobilize for change, one must actually move. Barry Hague and Brian Loader state that "democracy is about more than voting or providing better public information to the citizen. . . . Democracy has at its heart self-determination, participation, voice and autonomy."[59] Active participation is important because, as Jacques Ellul explains, political power comes through individual sacrifice: "If citizens are to experience power, there must be both personal devotion and time set aside for political activity."[60]

But dedication and mobilization alone do not a successful protest make, as illustrated by countless failed movements. The members of the movement must tread carefully and demonstrate considerable rhetorical sensitivity. Touraine argues that a social movement "can exist only if a collective action has social objectives, or in other words, recognizes that society has general interests or values. In other words, a social movement can exist only if collective action does not reduce political life to a confrontation between camps or classes, even though it organizes and extends conflicts."[61] It seems that the WNBR walks that balance between the general interest of the public and the differences between factions. The WNBR does not threaten drivers and resists an "us versus them" construction. By engaging in a humorous display of vulnerability, the protestors can gently point out the dangers inherent in bicycling in an urban environment while avoiding the appearance of pointing the finger.

Maintaining respect for the drivers is essential for several reasons. First, the participants can be certain that the event will be well attended by both participants and spectators alike. This is more than a protest action; this is an event. The organizers alert the news media and grant interviews. The press obliges and covers the event, alerting their readers and viewers. The public comes to witness the spectacle. Second, most of the observers are likely to identify more with car culture than bicycle culture. It would be difficult to keep the observers interested while employing a strident, self-righteous stance against car culture. Finally, the cyclists are operating from a position of reduced power. Despite the legal status that requires that cars share the road with cyclists, drivers will suffer little by ignoring this mandate while the cyclist would suffer greatly. Of course there would be legal repercussions for the driver, but the chances of death or severe injury are infinitesimal compared to those of the cyclist. The mass ride allows cyclists to take over the road and force the drivers to relinquish, if only temporarily, control of the road and yield to the cyclists without calling attention to this fact.[62] The rhetoric employed by the WNBR demonstrates that they recognize their fringe status within the dominant culture.

Like most protest actions, the WNBR is directed both internally and externally. Although bicyclists are a diverse lot, the WNBR attempts to create a collective identity for all of them based on vulnerability and exposure. This is necessary because there is little holding together such a disparate group. As Mary Bernstein writes, "Some sort of identity is necessary to translate individual interests to group interests and goals and individual agency to collective action."[63] This collective identity, according to William Gamson, is "manifested through the language and symbols by which it is publicly expressed—in styles of dress, language, demeanor, and discourse."[64] For bicyclists, the only symbol that they really seem to have in common is the bicycle. WNBR challenges that notion by providing them with another symbol—their relative nakedness among automobiles. Most bicyclists have had close calls with cars. Connecting the already existing feelings of caution in the presence of cars with the metaphor of nakedness as vulnerability and exposure allows organizers to portray naked protest as a natural expression of this shared experience.

Beyond the metaphorical arguments, what makes the WNBR particularly powerful is its non-institutional status. Like the lactivists, these individuals are engaging in direct citizenship. However, unlike PETA, which has an air of professionalization, professional activists do not seem to be a significant part of this movement.[65] This has implications for both the internal and external audiences. For the participants and sympathizers, there is the tacit understanding that there will be no others to step into that role, as in the case of other, more professionalized social movement organizations. As such, there is increased pressure to participate. This provides a more egalitarian structure to the event; one need only bring a bicycle, skateboard, rollerblades, or any other self-powered locomotion to participate. In other words, if this is something that the potential participant may be interested in, it is quite likely that he or she already has the means to participate. One need only disrobe and ride, and even that is optional with the mandate to "bare as you dare." If there is no ride in a particular city, people are free to simply organize one. According to the WNBR website, "WNBR is not an organization, it is a clothing optional bike ride to protest car culture. There is no head office or admin. It is about the creative empowerment of the individual. It is up to you to take ownership."[66] Removing possible barriers and objections to participation makes it clear that all potential participants are invited to ride. Moreover, this flat structure allows participants to direct the protest action as they see fit (within some provided guidelines) with no oversight from superiors.

For the observers, this near anarchic distribution of power has a different effect. Aristotle noted that for the audience to believe the speaker, one must *believe in* the speaker, calling this means of persuasion *ethos*.[67] Knowing that the participants are everyday people who have chosen to take part in this protest is a powerful argument for their sincerity. One expects professional activists to act in outlandish ways, but not accountants and schoolteachers. This invites the viewer to place themselves in the situation and consider how much they would have to believe in a cause to disrobe in public. Such a demonstration of sincerity functions enthymematically, with the audience supplying the evidence from their own discomfort with public nudity. The protesters need not argue their sincerity; the fact that they are naked induces the audience to make that argument for them.

Promises and Perils for the WNBR

WNBR is likely the largest nude protest in the world, with its protests spanning the globe. As such, it is worthy of close scrutiny to consider how the unclothed body can be employed in protest actions. WNBR illustrates the benefits as well as the pitfalls of such a protest. In fact, for many of the benefits, there is a corresponding problem that confronts the members of the WNBR.

From an instrumental perspective, the WNBR is a brilliant way to garner attention. Although some may consider the mode of protest odd or downright offensive, Anthony Downs explains that "the term *rational* is never applied to an agent's ends, but only to his means."[68] Downs elaborates that his usage of the term rational is that of a purely economic sense: "The economic definition [of rationality] refers solely to a man who moves toward his goals in a way which, to the best of his knowledge, uses the least possible input of scarce resources per unit of valued output."[69] If the goal is to increase awareness of the vulnerability of bicyclists on the road, the extensive media coverage demonstrates its effectiveness. Moreover, regardless of whether one agrees with all of the stated goals of the protest—protesting car culture and oil dependency and promoting body acceptance—the mode of nude protest on bicycles fulfills all of those ends. Yet this attention comes in the face of arguments that the WNBR is little more than a group of exhibitionist cyclists.

Secondly, WNBR uses nudity strategically to provide a powerful metaphor for exposure. This serves the dual purposes of forcefully directing the message to the plight of cyclists in the presence of cars and provides an umbrella capable of covering all potential cyclists thereby forming the basis of a collective identity for cyclists. However, this identity is far from empowering, claiming a victimized status. To subscribe to this collective identity is to surrender to fear. Yet the nudity serves to reinforce this vulnerability to both the participants and spectators alike.

Although the stated goal of promoting body freedom and acceptance is admirable, there are some dangers in employing this form of protest to those ends. Focusing on the body may individualize the protest in the minds of the participants. As stated before, social movements must help participants move away from the individual and toward the collective identity. There is the temptation to consider one's *own* body acceptance or freedom, losing sight of the general notion of body acceptance. This is in part because of the intensely personal nature of nudity. Public nudity is likely to induce a hyperawareness of oneself, making it difficult to lose oneself in the collective. Rather than participating as a member of a cohesive group, riders may feel as if they are isolated individuals riding along with other individuals. Instead of a collective action, it becomes the individual acts of many.

Finally, WNBR is a protest action that willingly embraces humor and fun. Phillip Carr-Gomm notes that "the atmosphere created is one of fun and of a summer party rather than one of 'manning the barricades' in a spirit of defiance and rebellion."[70] This party atmosphere functions to disarm a potentially hostile audience and provide an enjoyable experience for the participants. However, this pleasure can come at a cost. Lauren Langman argues that retreating into the "realms of privatized hedonism" can short circuit the potential for social change.[71] Moreover, their emphasis on fun runs the risk of trivializing the message of the protest. The cause of the bicyclists can be quite literally a matter of life and death. In 2009 there were 630 cyclists killed and 51,000 injured in collisions with automobiles.[72] Such statistics make clear the need for such a protest, but the jovial atmosphere can obscure this need.

Even with all of its weaknesses, WNBR demonstrates the potential for using the body in widespread global protest. The formula seems to have some merit considering that people in cities all over the world are taking part in this protest. Moreover, the protest seems to have staying power. Unlike the Facebook lactivist campaign discussed in the previous chapter which seemed to wither on the vine after the first year, WNBR is still going strong after eight years. This is particularly impressive in light of the lack of a formal structure, although this may be part of its strength. Perhaps this works because such a protest requires considerable investment in the cause; few can afford to be cavalier about biking naked through the streets of one's town. But once one joins, they seem likely to return, as some of the interviewees reported attending one or more previous WNBR events. Also, by decentralizing and allowing ordinary individuals to take the reins of the protest event, WNBR avoids potential bottlenecks and allows for considerable scalability. This reliance on everyday people concerned with the environment and the plight of cyclists provides an opportunity for ordinary citizens to do extraordinary things.

Notes

1. Sexual activity in automobiles has a long history: "While legislative inaction and parental trust enabled apparently sex-hungry teenagers to take to America's roads in the 1920s, the active state suppression of red-light districts pushed prostitutes too, into cars in secluded parking spots. Just as the moralists feared, the automobile had become a brothel on wheels." Peter Ling, "Sex and the Automobile in the Jazz Age," *History Today* 39, no. 11 (1989): 24.

2. "World Naked Bike Ride," Wikipedia, en.wikipedia.org/wiki/World_Naked_Bike_Ride.

3. Philip Carr-Gomm, *A Brief History of Nakedness* (London: Reaktion, 2010), 111.

4. World Naked Bike Ride, "World Naked Bike Ride (WNBR) - Naked Bicycle People Power! Stop Indecent Exposure to Vehicle Emissions!" www.worldnakedbikeride.org.

5. Rick Barry, "Oh, Sure: Now We're Having a Nude Bike-In," *Tampa Tribune* January 23, 2005, 1.

6. Hattie Garlick, "The Naked and the Dead Brave," *Times* (London), June 8, 2007, 2.

7. For more on trivializing social movements within the media, see Todd Gitlin, *The Whole World Is Watching: Mass Media in the Making & Unmaking of the New Left* (Berkeley: University of California Press, 1980), 27-28.

8. Moira Smith, "Humor, Unlaughter, and Boundary Maintenance," *Journal of American Folklore* 122, no. 484 (2009): 152.

9. Rashid Razaq, "Naked Cyclists to Stage London Demo," *Evening Standard* (London), June 5, 2006, 19.

10. "Take It Off," *Chicago Tribune* May 15, 2009, 36. This brief article provides descriptions of several events that encourage at least partial nudity including National Go Topless Day, Burning Man festival, the Naked Pumpkin Run, and Improv Everywhere's No Pants! Subway Ride.

11. Richard Chu, "Residents Walk, Run, Cycle Nude to Raise $150,000, Awareness," *Vancouver Sun* June 13, 2005, B1.

12. Catherine Saillant, "Ojai's Pastie Lady Takes Her Activism Elsewhere; She Heads to Oregon, Looking for Less Hassle and Greener Mind-Sets," *Los Angeles Times* May 11, 2008, B3; Catherine Saillant, "She's Got Ojai's Attention; an Activist Who Bikes through Town in Only Pasties and a G-String Has Supporters and Critics. An Easter Incident Could Lead to Charges," *Los Angeles Times* April 26, 2008, B1.

13. "Bicyclists Ride in Protest, and in Little Else," *New York Times* June 12, 2005, 1.9.

14. Susan Mohammad, "Protesters Grin and Bare It Despite Chilly Conditions," *Ottawa Citizen* June 11, 2006, A11.

15. Diane Toroian Keaggy, "Cyclists Will Get Naked For . . . Does It Matter? St. Louis Will Take Part in World Naked Bike Ride on Saturday, Calling It a Stand against Oil Dependence," *St. Louis Post-Dispatch* July 31, 2008, C4.

16. "Take Off Your Car and Join the Nude Cyclists," *Ottawa Citizen* December 26, 2007, A17.

17. Victoria Davis, "Safety for Cyclists," *Nelson Mail* (Nelson, NZ), March 1, 2007, 9.

18. Barry, "Oh, Sure," 1.

19. Jeremy Schwartz, "Cheeky Cyclists to Cruise Austin in Clothing-Free Ride," *Austin American Statesman* June 11, 2004, A1.

20. Colette Bernhardt, "The Guide: Events: The Guide: World Naked Bike Ride Brighton, Cardiff, London, Sheffield," *The Guardian* June 13, 2009, 42.

21. Clare Trapasso and Jotham Sederstrom, "Bikers Strip for Rally vs. Cars and Oil," *New York Daily News* June 14, 2009, 2.

22. Matt Gleason, "'Naked' Ride Will Find Pedalers in Skivvies," *Tulsa World* July 23, 2009, D3.

23. Louise Roseingrave, "Wheely Brave," *Daily Mirror* (London), July 20, 2010, 3.

24. See Gavin Brown and Jenny Pickerill, "Space for Emotion in the Spaces of Activism," *Emotion, Space and Society* 2, no. 1 (2009): 24-35.

25. See Steven E. Barkan, Steven F. Cohn, and William H. Whitaker, "Beyond Recruitment: Predictors of Differential Participation in a National Antihunger Organization," *Sociological Forum* 10, no. 1 (1995): 113-34; Bert Klandermans and Dirk Oegema, "Potentials, Networks, Motivations, and Barriers: Steps Towards Participation in Social Movements," *American Sociological Review* 52, no. 4 (1987): 519-31; Herbert W. Simons, "Requirements, Problems, and Strategies: A Theory of Persuasion for Social Movements," *Quarterly Journal of Speech* 56, no. 1 (1970): 1-11.

26. Simon Cottle, "Reporting Demonstrations: The Changing Media Politics of Dissent," *Media, Culture & Society* 30, no. 6 (2008): 853.

27. Al Kratina, "Baring It All to Bring About Change; Nudity Is Optional—But Encouraged—At Today's World Naked Bike Ride," *The Gazette* (Montreal), June 12, 2010, F2.

28. Keaggy, "Cyclists Will Get Naked," C4.

29. Mohammad, "Protesters Grin and Bare It," A11.

30. Joel Currier, "Mostly Naked Cyclists Draw Crowd for Ride," *St. Louis Post-Dispatch* August 3, 2008, C2.

31. "Cyclists to Give Their Cause Extra Exposure," *Daily Mail* (London), June 12, 2004, 4.

32. "Up to 200 Naked Cyclists Expected on City's Streets," *Irish Examiner* (Cork, Ireland), July 19, 2010.

33. Gale Hayley, "Bare Biking a Real Sea Change," *Nelson Mail* (Nelson, NZ), March 15, 2010, 3.

34. "Bicyclists Ride in Protest, and in Little Else," 1.9.

35. Paul Moloney, "A Wheely Cheeky Protest; World Naked Bike Ride Cyclists Doff Duds to Promote 2-Wheeled Solution to Air Pollution," *Toronto Star* June 10, 2007, A2.

36. Hayley, "Bare Biking a Real Sea Change," 3.

37. Gleason, "'Naked' Ride Will Find Pedalers in Skivvies," D3.

38. Gleason, "'Naked' Ride Will Find Pedalers in Skivvies," D3.

39. See John B. F. De Wit, Enny Das, and Raymond Vet, "What Works Best: Objective Statistics or a Personal Testimonial? An Assessment of the Persuasive Effects of Different Types of Message Evidence on Risk Perception," *Health Psychology* 27, no. 1 (2008): 110-15.

40. De Wit, Das, and Vet, "What Works Best," 113.

41. Julie Pace, "Protestors Shun Bike Shorts to Bare Views on Oil, Cars," *Tampa Tribune* March 13, 2006, 6.

42. Quoted in Shelby Martin, "Cyclists in Bike Parade to Air Views in Ride of Protest," *Register-Guard* (Eugene, OR), March 14, 2009, A1.

43. Eric Lenkowitz, "Weird but True," *New York Post* June 10, 2007, 23.

44. Smith, "Humor, Unlaughter, and Boundary Maintenance," 152.

45. Jeffrey S. Juris, "Performing Politics: Image, Embodiment, and Affective Solidarity During Anti-Corporate Globalization Protests," *Ethnography* 9, no. 1 (2008): 74.

46. James Tait, "Buff Riders Saddle the Blame," *Daily Mirror* (London), June 12, 2004, 27.

47. Thomas W. Krause, "For Safety, Helmets; for Decency, G-Strings," *Tampa Tribune* March 11, 2006, 5.

48. Krause, "For Safety, Helmets," 5.

49. For discussion of some legal aspects of nudity, see Libby S. Adler, "A Short Essay on the Baring of Breasts," *Harvard Women's Law Journal* 23 (2000): 219-26; Anita L. Allen, "Disrobed: The Constitution of Modesty," *Villanova Law Review* 51, no. 4 (2006): 841-58; George J. Eigenhauser, Jr., "Criminal Sanctions against Nude Sunbathing in California," *Criminal Justice Journal* 1, no. 3 (1978): 491-508; Leonard J. Michaud, "Decent Exposure: An Inquiry into the Constitutionality of State Statutes Proscribing Expressive Public Nudity Comment," *Loyola Law Review* 22 (1975): 1018-35; Helen Pundurs, "Public Exposure of the Female Breast: Obscene and Immoral or Free and Equal," *In the Public Interest* 14 (1994-1995): 1-38. There have been many cases regarding the Constitutional protection of nude dancing in particular, only a few of which are cited here. Legal scholar Amy Adler notes that the image of the nude dancer has been problematic for the courts because it threatens "the fantasy of a coherent First Amendment discourse." Indeed, Alan Howard observes that in the case of *Barnes v. Glen Theatre, Inc.*, "the five justice majority needed three separate opinions and three different rationales to explain why the public nudity ban could be applied constitutionally to nude

dancing." Clearly, the nature of nudity is still controversial and difficult to resolve. Amy Adler, "Symptomatic Cases: Hysteria in the Supreme Court's Nude Dancing Decisions," *American Imago* 64, no. 3 (2007): 313; Alan J. Howard, "When Can the Moral Majority Rule? The Real Dilemma at the Core of the Nude Dancing Cases," *Saint Louis University Law Journal* 44, no. 3 (2000): 897.

50. Danielle Moriber, "A Right to Bare All? Female Public Toplessness and Dealing with the Laws That Prohibit," *Cardozo Public Law, Policy, and Ethics Journal* 8, no. 2 (2010): 473.

51. "Cyclists to Give Their Cause Extra Exposure," 4.

52. For example, in the case of *Erie v. Paps A.M.*, which considered the constitutionality of requiring nude dancers to wear pasties and g-strings rather than appearing completely nude, the Supreme Court argued that "the ordinance does not attempt to regulate the primary effects of the expression, i.e., the effect on the audience of watching nude erotic dancing, but rather the secondary effects, such as the impacts on public health, safety, and welfare, which we have previously recognized are 'caused by the presence of even one such' establishment." *Erie v. Paps A.M.*, 529 U.S. at 291 (2000). Yet such a stance is not without its dissenters, even within the court. Justice Souter notes in his opinion: "It is not apparent to me as a matter of common sense that establishments featuring dancers with pasties and G-strings will differ markedly in their effects on neighborhoods from those whose dancers are nude. If the plurality does find it apparent, we may have to agree to disagree." In other words, the argument that banning nude dancing while replacing it with similar erotic expression (but covered with pasties and a g-string) will remedy the secondary effects seem as flimsy as the prescribed dancer coverings. See *Erie v. Paps A.M.*, 313 n.2.

53. Lloyd F. Bitzer, "Aristotle's Enthymeme Revisited," *Quarterly Journal of Speech* 45, no. 4 (1959): 408.

54. John Ince, *The Politics of Lust* (Amherst, NY: Prometheus Books, 2005), 32.

55. Michael Calvin McGee, "'Social Movement': Phenomenon or Meaning?" *Central States Speech Journal* 31 (1980): 233. This is quite similar to McCarthy and Zald's definition of a social movement as "a set of opinions and beliefs in a population which represents preferences for changing some elements of the social structure and/or reward distribution of a society." John D. McCarthy and Mayer N. Zald, "Resource Mobilization and Social Movements: A Partial Theory," *American Journal of Sociology* 82, no. 6 (1977): 1217-18.

56. Arthur W. Frank III, "Reality Construction in Interaction," *Annual Review of Sociology* 5 (1979): 167.

57. Alberto Melucci, *Challenging Codes: Collective Action in the Information Age* (Cambridge: Cambridge University Press, 1996), 14.

58. Alain Touraine, *What Is Democracy?* trans. David Macey (Boulder, CO: Westview Press, 1997), 57.

59. Barry N. Hague and Brian Loader, "Digital Democracy: An Introduction," in *Digital Democracy: Discourse and Decision Making in the Information Age*, ed. Barry N. Hague and Brian Loader, 3-22 (London: Routledge, 1999), 7.

60. Jacques Ellul, "Technology and Democracy," in *Democracy in a Technological Society*, ed. Langdon Winner, 35-50 (Dordrecht: Kluwer, 1992), 42.

61. Touraine, *What Is Democracy?* 57.

62. Even then, sometimes drivers refuse to give up this power. Anne Saker reports on a 2006 WNBR in Portland, Oregon in which a participant directing traffic states that a driver "honked her horn, yelled at him and lurched her SUV forward," knocking him to

the ground. The participant, Phil Sano, was arrested and charged with disorderly conduct, criminal mischief and indecent exposure, but the charges were later dropped. There was no mention of the driver having been charged, although the article did note that the driver's boyfriend was an off-duty police officer. Portland is a particularly bicycle friendly city with lax public nudity laws and robust protection of freedom of expression, so this case stands out all the more. Anne Saker, "Cyclist Was Ready to Lay Bare Case—Until It Was Dismissed/Minister Was Arrested on World Naked Bike Ride, but Doesn't Need His Day in Court," *Houston Chronicle* July 1, 2007, 6. For more on Oregon law concerning public exposure, see Thomas P. Harbolt, "Welcome to Oregon! No 'Pasties' Or 'G-Strings' Required: *Barnes v. Glen Theater, Inc.*," *Willamette Law Review* 29, no. n1 (1993): 147-49.

63. Mary Bernstein, "The Analytic Dimensions of Identity: A Political Identity Framework," in *Identity Work in Social Movements*, ed. Jo Reger, Daniel J. Myers, and Rachel L. Einwohner, 277-301 (Minneapolis: University of Minnesota Press, 2008), 278.

64. William A. Gamson, "Constructing Social Protest," in *Social Movements and Culture*, ed. Hank Johnston and Bert Klandermans, 85-106 (Minneapolis: University of Minnesota Press, 1995), 100.

65. For more on the professionalization of social movements, see Kevin Djo Everett, "Professionalization and Protest: Changes in the Social Movement Sector, 1961-1983," *Social Forces* 70, no. 4 (1992): 957-75; John D. McCarthy and Mayer N. Zald, "The Trend of Social Movements in America: Professionalization and Resource Mobilization," in *Social Movements in an Organizational Society: Collected Essays*, ed. Mayer N. Zald and John D. McCarthy, 337-91 (New Brunswick, NJ: Transaction Books, 1987); Suzanne Staggenborg, "The Consequences of Professionalization and Formalization in the Pro-Choice Movement," *American Sociological Review* 53, no. 4 (1988): 585-605.

66. World Naked Bike Ride, "Organizing a Ride," wiki.worldnakedbikeride.org /index.php?title=Organizing_a_ride.

67. Aristotle, "Rhetoric," in *The Complete Works of Aristotle: The Revised Oxford Translation*, ed. Jonathan Barnes, 2152-269 (Princeton, NJ: Princeton University Press, 1984), 1356a2-14, 1378a7-20. Cicero likewise notes the importance of sincerity on the part of the speaker, explaining that when trying to elicit an emotion in an audience, he or she must also genuinely feel such emotion: "It is impossible for the listener to feel [emotions] . . . unless all those emotions, which the advocate would inspire in the arbitrator, are visibly stamped or rather branded on the advocate himself." Marcus Tullius Cicero, *De Oratore, Books I-II*, trans. E. W. Sutton and H. Rackham (Cambridge, MA: Harvard University Press, 1942), II.XLV.189. For more on how ethos functions physiologically and psychologically, see Brett Lunceford, "The Science of Orality: Implications for Rhetorical Theory," *Review of Communication* 7, no. 1 (2007): 95-97.

68. Anthony Downs, *An Economic Theory of Democracy* (New York: Harper, 1957), 5.

69. Downs, *Economic Theory of Democracy*, 5.

70. Carr-Gomm, *A Brief History of Nakedness*, 112.

71. Lauren Langman, "Culture, Identity and Hegemony: The Body in a Global Age," *Current Sociology* 51, no. 3-4 (2003): 245.

72. NHTSA's National Center for Statistics and Analysis, "Traffic Safety Facts 2009 Data: Bicyclists and Other Cyclists" (Washington, DC: National Highway Traffic Safety Administration).

Chapter 5
Political Display and the Mediated Body: Exhibition and Politics on CollegeHumor.com

One can sometimes find politics in the strangest of places. Collegehumor.com is hardly the place that one would expect to find political action. After all, this is a site dedicated more to images of passed out drunk college students who have been vandalized by their friends through the liberal use of a Sharpie pen. On this site it is hardly shocking to find nudity; in fact they often encourage sending in such photographs. But even in this unlikely place one can find political action. During the 2006 presidential campaign between John Kerry, George W. Bush, and, to a lesser extent, third party candidates such as Ralph Nader, College Humor ran a "contest" in which females were encouraged to write the name of their preferred presidential candidate on their naked bodies and submit a photograph to the website.[1]

This contest draws on a long tradition of nudity and authenticity. John Pratt notes that pornographic magazines have long recognized the power of normalizing sex and removing it from exotic locales and the world of fantasy, describing *Fiesta*, a British magazine, that featured "female bodies displayed and situated in the world of launderettes and similar mundane places."[2] Ruth Barcan likewise notes that "in recent years new genres of images of 'ordinary people' naked have begun to circulate in both the private and public domains. Such images can be glamorous or they can be raw. On the one hand, there are imitations of glamorous celebrity nudity—for example in the form of boudoir photography. At the other end of the scale is a range of practices in which ordinary people, pictured nude, are celebrated precisely for their ordinariness."[3] The amateur status of the College Humor participants provides a sense of authenticity for both the images and the political views that the women represent. These are real people and not simply poll numbers. Perhaps the exposure of one's body provides a way to incorporate the body back into politics. But the fact that the vast majority of these women do not show their faces (in other words, their identity) creates not a disembodied polis, but rather a partially embodied polis that is sexualized and cognizant of that power. This is objectification in the service of politics, and objectification is brought to the forefront.

The images themselves largely support either George W. Bush or John Kerry, although there are a handful of Nader (independent) and Badnarik (Libertarian) supporters. In the instances of nude protest I have examined thus far, the causes tend to lean toward the liberal end of the spectrum, so it is interesting to note that in this case there are a significant amount of Bush supporters.[4] In this mode of expression the arguments are literally inscribed upon the body[5] and there is a kind of exhibitionistic anonymity in that many of them are depicted without showing their faces. The medium allows each of these women to effectively manage their presentations of self and the use of nudity as part of that presentation is a powerful rhetorical tool because of its ability to induce emotion in the viewer.

In these images, the women are visible as an embodied entity, yet only present as an ephemeral image. Their nakedness will not allow the viewer to consider them as disembodied entities; as such, these images remain inextricably linked to the embodied women. But this tenuous link to the body, especially in light of the choice a large majority of them made to remain anonymous, makes it difficult to consider exactly *whose* body we should link the image to. This too can be a powerful rhetorical strategy. Roland Barthes writes, "Since every photograph is contingent (and thereby outside of meaning), photography cannot signify (aim at generality) except by assuming a mask."[6] Perhaps this is one way that the women in the College Humor survey were able to convey their meaning as a more generalizable statement of womanhood voting. By not revealing their faces, they mask their identities while disclosing their feminine attributes. In this way, they are each not merely an instantiation of a particular individual woman—they transcend their individuality to become instantiations of a universal woman. Each woman stands synecdochically in place of all women.

Although such displays may seem trivial when compared to more organized protest actions such as the World Naked Bike Ride or those performed by PETA, these images provide insights into how individuals perceive their own bodies as they place them into the public sphere. In other words, there are specific strategies that these women see as open to them and useful in achieving these ends. By examining these vernacular displays of political nudity, one can gain a sense of how ordinary individuals (i.e., not activists) may view the display of the body. Finally, such displays demonstrate the variety of means by which individuals can work for political ends. As Robert Asen writes, "Rather than asking what counts as citizenship, we should ask: how do people enact citizenship? Reorienting our framework from a question of *what* to a question of *how* usefully redirects our attention from acts to action. Inquiring into the *how* of citizenship recognizes citizenship as a process. From this perspective, citizenship does not appear in specific acts per se, but signals a process that may encompass a number of different activities."[7] In considering these images as performances of citizenship rather than as exhibitionism one may obtain a small glimpse into the fringes of political behavior and gain insights into how the body functions politically. As Gerard Hauser observes, "The body, as a corporeal entity, is an organism; its biological status is not symbolic. At the same time, limiting our understanding to its status as a biological organism ignores the body's symbolic significance and the numerous ways in which it is used as a form of signification."[8] This chapter will examine some of those ways.

Examining the Images

Roland Barthes suggests that "all images are polysemous; they imply, underlying their signifiers, a 'floating chain' of signifieds, the reader able to choose some and ignore others."[9] The solution to this infinite range of ambiguity is for societies to prescribe specific ways of seeing particular images. Western society has become accustomed to the use of sensual female imagery in the media and has coded these displays as sexual enticement. However, this is further defined as a kind of passive enticement. Women (and adolescent girls) are not *supposed* to have sexual desires.[10] Such norms are reinforced through the media. In their study of women's magazines *Cosmopolitan* and *Cleo*, Panteá Farvid and Virginia Braun found that "women were not constructed as 'inherently' sexual in the way that men were. Rather, female sexuality was (implicitly) constructed as 'catching up' to (an ever present and pre-existent) male sexuality, which ostensibly constitutes 'real' sexuality."[11]

Bahar Hanjani writes, "Flipping through an American magazine, one realizes that Americans are infatuated by female sexuality—the images of perfect beauty. Almost every conceivable product has a beautiful woman selling it. There is an emphasis on the desirability and sexiness of women in ads."[12] But such sentiments can be found even in cultures that do not embrace the "sex sells" mentality of the United States. Hanjani compared Iranian and American advertising and found that even without the presence of female models, "the advertisements of women's products in Iran reveals the influence of Western rhetoric. As in the American ads, Iranians (in an ambiguous way) communicate the significance of a woman's beauty in her life."[13] In advertising, images of female nudity seem increasingly common.[14] As with advertising, one is likewise confronted with images of the disrobed female body on television shows.[15]

In stark contrast to print and broadcast media, images of nudity on the internet are much less regulated and quite easy to find. Even so, such images have the potential to enact social protest. In his discussion of mediated exhibitionism, Matthew Jones argues that "in a limited way, nudity represents a threat to the established order because it denies the naturalization of that order and reveals the potential for alternatives that become visible when social roles and bureaucracy are 'stripped off.'"[16] One remarkable example of this can be seen in the case of Aliaa Magda Elmahdy, an Egyptian blogger who caused an uproar by posting nude photographs of herself online. The *New York Times* reports that "Ms. Elmahdy—whose boyfriend, Kareem Amer, spent four years in jail for writings deemed insulting to Islam and Mr. Mubarak—posted the photographs with a statement declaring them an act of rebellion against Egypt's conservative culture and 'sexual complexes,' in the spirit of the revolution."[17] Thus the use of nudity, even online, can send a strong message.

As discussed in the chapter on lactivism, the Internet provides a way to combine private and public space in ways previously unimaginable. The College Humor participants are acting in the public sphere from the privacy of their homes, displaying an intimate view of their bodies for public consumption. But these images are not the pseudointimate depictions of pornography or amateur webcams; rather these women combine the intimate and the political in order to stand and be counted. These women construct displays of political advocacy through one of the most personal means possible—the intimate display of their bodies. This section examines how these women rhetorically constructed their messages using the following themes: insults against the candidates and the other participants; visual arguments; body exhibition; exhibition of identity through showing the face; prostitution and sexual power; arguments concerning values; and arguments concerning other candidates.

Insults

The use of insults is prevalent in these images. Although some of the messages in both the captions and the bodies focus on arguments, many are reduced to slogans or simple ad hominem attacks. For Bush, many of the messages focus on his intelligence. For example, three different women hold bumper stickers with similar messages: "It takes a village idiot," "Somewhere in Texas there's a village missing an idiot," and a picture of Bush with "American Errorist." One Bush supporter provided a kind of backhanded compliment to Bush. The caption above her photo states: "I love BUSH !!!!!! No matter how big of an idoit [sic] he is, anyone is better then Kerry." Those who opposed Kerry tended to focus on his tendency to "flip flop." One Bush supporter is shown standing in the shower holding her breasts in her hands. The caption reads "VOTE BUSH! NO FLIP FLOPPIN HERE!" Another, more visual depiction has the caption "I Pledge Allegiance to the Flag of Bush Country!" Her hand is over her left breast and she has a sticker on that hand that states "studentsforbush.com no flip flops!" with a circle and slash motif over a picture of a pair of flip flops.

Several of the images resort to name-calling. Two separate Bush supporters state that "Kerry is a bitch" or a "little bitch." A Kerry supporter has "Bush is a boob" written on one of her breasts. Another Kerry supporter has "Bush is a bum Vote Kerry" written across her buttocks. A crude, yet calculated ad hominem attack is found in the image of a Kerry supporter who has "Vote Kerry" written on her breasts and just above her panty line is written "Bush is a Fag!" The placement of this particular attack seems intentional. By casting aspersions on Bush's sexuality near the location of her sexual organs, she places it within view of those who would notice—those who would be interested in that portion of her covered anatomy. At the same time, she places it in a location that would be of little interest to a homosexual man. A similar argument is made by a woman who attacks both candidates. "Come to Mommy" is written between her breasts, with "Bush" written on her left breast and "Kerry" written on the right. Beneath the chain that connects her nipples is written "Chain of Fools – Any real men running?" A stick figure is drawn coming out of her panties over her pubic area.

The insults do not stop with the candidates, but spill over into insults against the supporters. The image of one Bush supporter carries the caption "Why be an ASS with a rack like this? From the land of the rising sun: an absentee-voting exchange student for Bush in 2004." Yet despite her obvious pride in her breasts, she is unwilling to display them; her breasts are concealed by her hair and a bumper sticker. The caption for another Bush supporter asks, "Why must all the Kerry supporters have pepperoni nipples?? (Nudity..not pepperonis!!)," engaging not only in an attack on the bodies of those who support the other candidate, but also a kind of depersonalization.

By reducing the opposition to inanimate objects, one need no longer engage their arguments. Haig Bosmajian suggests that "one of the first important acts of an oppressor is to redefine the oppressed victims he intends to jail or eradicate so that they will be looked upon as creatures warranting suppression and annihilation. I say 'creatures' because the redefinition usually implies a dehumanization of the individual."[18] Perhaps the online world makes it easier to dehumanize others because they are not present. Leopoldina Fortunati argues that because of our reliance on communication technologies, "beyond the remaining old poverty, which exists even in the industrialized nations, the new poverty that affects everybody is a *poverty of first-hand reality*."[19] When the image is just an image rather than a person, it is much easier to disconnect the icon of the individual from the individual themselves and forget that the image is merely a representation of a person who has a body, emotions, and value. This tendency to dehumanize others is dangerous, whether one is a candidate or just a supporter. The reduction of Bush to his namesake (both vulgar and common) is another form of dehumanization that I will discuss further in the next section.

Visual Arguments

Valerie Peterson argues that in today's media-rich environment, "the balance of power has shifted between words and images."[20] In a contest such as this where the body itself can function as an argument, it is unsurprising that there are several instances of visual argument in the images. The most obvious one is the relationship between Bush's name and the common term for a woman's pubic hair. In Barbara Phillips and Edward McQuarrie's taxonomy of visual metaphors, such a strategy falls into the category of *replacement*, in which the image of the woman's pubic hair (bush) stands in for the non-present image of the President (Bush).[21] Many women made this observation, both for and against Bush. Several Bush supporters have arrows pointing from their exposed breasts down to their pubic areas, with various levels of exposure. For example, one image is an extreme close-up of the woman's pubic mound with the word "Bush" written directly above it. Another has a photograph taken from the reclining woman's point of view looking down; above her pubic area is a piece of paper that states "Bush is the only way to go! Vote Bush!" Her pubic area, trimmed in the "landing strip" style, is visible beneath the paper. Most of the Bush supporters with exposed pubic hair seem to have natural pubic hair, so this image of a trimmed pubic mound seems out of place. The landing strip style of pubic hair is sometimes referred to as the "Playboy" bikini wax because it is a common style found on *Playboy* centerfolds.[22] Thus a woman who supports the Republican candidate—a party that often opposes pornography—seems to have internalized the pornographic aesthetic. However, in accordance with the traditional stance espoused by the Republican party, most pro-Bush women are more demure. One Bush supporter has Bush written on her abdomen. Her hand touches her partially exposed pubic area, further reinforcing the message. The caption of another image makes the link between femininity and Bush more explicit, stating, "This is the way all girls should vote." The woman in the picture has "Bush 04" written across her exposed breasts and a large arrow pointing to her (non-exposed) pubic area. Others have similar messages with arrows pointing toward their pubic area, which is either cropped out of the picture or covered with panties.

Kerry supporters also draw on the Bush/pubic hair theme, but with a call to remove Bush, making the argument with their bodies as well as their words. One Kerry supporter has several messages inscribed upon her body. "Vote Kerry" is written across her breasts, "Vote Nov 2" is written across her belly button, and "No Bush!" is written across her lower abdomen with an arrow pointing to a completely shaved pubic area. Other images have similar messages such as "Say no to Bush" and "No Bush" over their exposed, shaved pubic areas. Several images explicitly state that voters must "shave the bush." In one image, the woman holds a razor over the words "no more bush," written on a partially exposed pubic area that had been shaved. The sudden removal of Bush by way of the blade functions as an act of symbolic violence; women have the power to cut the representation of Bush out of the body politic. Such an image also seems to function as an index of desired action. Semiotician Charles Peirce explains that "An *Index* is a sign which refers to the Object that it denotes by virtue of really being affected by that Object."[23] The typical explanation of an indexical sign is that of a thermometer; as the temperature rises, the mercury rises. In this case, these women manipulate their bodies to augment their message such that their physical appearance represents the ultimate goal of the Kerry supporters—the removal of Bush. One Kerry supporter has "no bush 04" written directly onto her shaved pubic area. Another Kerry supporter echoes this with the caption over her picture: "Get rid of that Bush once and for all, vote Kerry to Shave America!" She is wearing panties with cutouts—in the holes are the words "Bush Sucks" and on each thigh, "shave" and "America." Following the same trajectory, another Kerry supporter creates a slightly dissonant message with her image. The caption states, "Trim BUSH in 04!" Below her navel is the word "Vote" pointing to a Kerry sticker. The word "Eat" has an arrow pointing to her pubic area. Unlike others who removed their pubic hair, this woman provides a different kind of symbolism. The ambiguity of the word "eat" allows for interpretations of consumption and the sexual act of cunnilingus, and perhaps both—Bush/bush can be overcome or conquered through acts that facilitate female sexual pleasure.

Another visual argument in these images is the idea of weapons of mass destruction or military power. Both Bush and Kerry supporters draw parallels between the idea of weapons of mass destruction and their breasts. A Bush supporter has "Weapons of Mass Destruction" and "Bush" written on her breasts, echoing the rationale for war in Iraq. A Kerry supporter has written on her breasts, "Only John Kerry can have these weapons of mass destruction." The caption reads, "My vote is for weapons of mass destruction." This paints the theme of weapons of mass destruction in a different light, suggesting the idea of female sexuality as a potentially destructive force. More importantly, her caption is a call for a feminine version of weapons of mass destruction, one based on female sexuality, which has often disrupted masculine politics in various ways. Another Kerry supporter also makes the case for breasts (feminine) rather than bombs (masculine power) by writing "Vote Kerry Boobs not Bombs" between her breasts with arrows pointing to her breasts.

Bush supporters seem to feel that the militaristic principles embodied in Bush are the ideal. A striking example of this is found in a Bush supporter whose caption reads, "Your Military Needs President Bush!!!" In the picture, she wears a green flight jacket with a patch that reads "Bush '04." She also wears beads with military men on them that go between her breasts. There is an American flag in the background. The image evokes the military persona of Bush in the flight suit and the "Mission Accomplished" banner on the aircraft carrier USS *Abraham Lincoln*. This is a striking image with rich detail, but the aspect I wish to examine further is the string of soldier beads that cascade between her breasts. That she would adorn herself with these toylike soldiers suggests a sense of ownership. There is no indication in the image or in the caption that she personally knows a soldier or has their interests at heart. Rather than display the image of, or make reference to, an actual soldier, she brings toy soldiers into the frame. By avoiding the harsh realities of death and injury that come with real soldiers, she sustains an illusion that all is well with the plastic, whole soldiers. But the commander in chief does not send plastic bodies to war. Rather, the bodies that he sends are connected to others and subject to destruction; plastic soldiers have no families and can be mass produced. In this image, it seems that this Bush supporter displays a disregard for the realities of war, viewing the soldiers as simply another resource to deploy—another piece in a game.

Two further examples of visual argument concerning clothing warrant discussion. A Badnarik supporter is shown wearing a bra with "Badnarik '04" written over her breasts. The caption states, "Vote libertarian . . . we have never stuffed our bras or ballots—LP.org." In this case, she subtly attacks both the bodies and the political parties of the other supporters. Her party is the party of fairness. But in light of the caption, her party is also the party of real women. The Libertarian woman has no need of artifice to enhance her beauty. In another political use of clothing, a Kerry supporter is depicted with exposed breasts but wearing what appears to be a gay pride flag from the waist down. Both hands are forming the V peace sign. "Mt. Holyoke 4 Kerry" is written on her upper torso. There is no mention of homosexuality or affiliation with a homosexual group. Even so, the flag serves as a kind of code for those who would recognize the symbolism and excludes those who would not.

Exhibition

Although there are many instances of political speech, some women seem to view this contest mainly as an excuse to expose their bodies. Even so, despite the implied (or expressed) desire to expose themselves, they still do not expose their faces. In fact, only a few individuals (male and female) expose their faces in this contest (these individuals will be discussed in the next section). One Kerry supporter is rather frank about her motives: the caption above her photo states, "Exhibitionist for Kerry." Another Kerry supporter has the caption, "My presidential contribution. With a nipple ring." She has "Kerry" written above her breasts, but the way she is turned in the frame brings the pierced breast to the foreground, drawing attention to her nipple piercing. Thus, her display of her nipple piercing is equally prominent to her display of political affiliation. Perhaps it is a function of nipple piercings to wish them displayed, because such blatant exhibitionism can also be found among Bush supporters. One Bush supporter writes, "I'll vote for bush, but I really just want to show off my piercings ;)." As with the Kerry supporter, the piercing to be displayed is turned such that it occupies the foreground of the image (her other breast also has a visible, albeit less elaborate piercing). She has a small American flag painted onto her breast, with her nipple taking up most of the blue field. Her nipple has a barbell with a star hanging on it. The piercing fits the flag motif, providing unity to the display, but, like the Kerry supporter, the main impulse seems to be a desire to display her piercing.

In contrast to those who expose their bodies for the pleasure of doing so are those who appear to use the forum as a means of gaining approval and esteem. One striking example of this impulse is found in the caption of a Bush supporter: "**please take this picture! MY BOOBS FOR BUSH!!! . . . are they okay?" Another Bush supporter demonstrates similar feelings in her caption: "I know the quality of this pic sucks but I get $20 if its posted and plus I did the BUSH sign all by myself so make me feel special PLEASE." "Bush" is written on her abdomen, below her bare breasts. The ability to feel "special" seems to require external validation for these women. Rather than wishing to expose themselves for the pleasure of being seen, these women use self exposure as a way to gain increased self-esteem through positive feedback. However, it is unclear how such feedback is to be received. Perhaps the act of posting the picture by the gatekeepers at College Humor deems them worthy of exposure to a larger public, allowing these women to feel special or desirable.

These women actively seek the gaze and demand to be viewed, but the gatekeepers have the final say in the matter. As Jones observes, "social forums that are permissive of public nudity often circumscribe it to approved body types and, in so doing, transform the naked body into an unnatural costume."[24] Some participants reveal that they had previously been rejected. For example, a Kerry supporter's caption states, "Got one of my boobs for the election erection contest.. I've sent you so many pictures and you never post them! Please post this one!!" Another Kerry supporter expresses a similar sentiment in her caption: "I've participated in the contest... now will you accept my photo?" She holds a sign that says "More trees Less Bush! I'm for Kerry!" The sign also has "(heart symbol) CH.com" written on it, perhaps to increase her chances of being posted by appealing to the site administrators' vanity. Thus it seems that rejection is a matter of consequence for these women, a signal that they were not worthy of having their bodies displayed that compels them to beg to be objectified. Yet with this objectification comes validation; she is worthy of being considered a sexual object. This may be an internalized response to the everyday objectification of women in Western society. To be worth something, a woman must live up to an ideal of beauty; in short, she must be worth seeing to be worth anything. As Iris Marion Young argues, women "experience our position as established and fixed by a subject who stands afar, who has looked and made his judgment before he even makes me aware of his admiration or disgust."[25]

Bridging these two extremes between sheer joy of exposure and the desperate seeking of external approval is the Kerry supporter with the caption, "Naked Campaign Contest. Please put my boobies up!" "Vote" is written above her breasts and "Kerry" is written below. It is as if she did not wish to mar the true attraction of the picture—her breasts—with a marker. She points to herself with both hands, one finger toward the word "Kerry," the other toward her left breast. Her face is clearly displayed, removing her anonymity. Not only does she display her breasts, but she willingly identifies herself as their owner and possessor. But on her face is an expression of questioning pleading rather than the expected confidence of one who would display herself in such a way. Her head is cocked to the side, her brow is furrowed, her mouth only half-smiling. Her attention to her body is apparent through details such as the French manicure on her fingernails and the well-applied makeup. Although she presents herself as someone who wishes to be exposed and as one who has pride in her body, she seems to lack the confidence of one who knows that she is desirable.

Exhibition of Identity: Faces

Most of the participants took advantage of the relative anonymity afforded by the medium and either cropped the photo or digitally manipulated it in order to keep their faces hidden. Only seven images out of 250 feature unobscured faces. All but one are Kerry supporters. Bush supporters may have more incentive to obscure their identity than avowed liberals. As Jones explains, "The face is a marker that designates ownership of a body and its actions to a particular self."[26] Refusing to reveal one's face allows her to likewise hide the action itself. The sole Bush supporter has the caption "Serena from www.stunningserena.com for Bush 2004!" Her whole body is visible, including her face, but she covers her breasts with her arm and wears a Bush sign around her waist. However, this seems like merely another way for her to promote her website, which requires a membership fee. Even so, it seems peculiar that the purveyor of an online pornography site would support Bush. In fact, it is surprising that there are a significant amount of Bush supporters disrobing at all. Bush represents the party least likely to embrace such exhibition; Republicans have often worked to ban the very images that these women display in support of Bush. The caption of a Kerry supporter clearly illustrates the irony in linking online nudity with support for Bush: "You Bushies realize that Bush is anti-porn, right? HELLO! Kerry all the way!" In the image she has "Bush is against online nudity. I'm for Kerry" written on her breasts.

Perhaps the idea of owning one's sexuality and body is more consistent with liberal ideals than conservative. For a Bush supporter to take part in such a political display is to use the tactics of the "other" side; taboos against public displays of nudity and online pornography must be broken if the Bush supporters are to play the same game as the Kerry supporters. One Bush supporter states in her caption: "Pic for the political boobs contest. Sick of seein so many libs on here. Bring on the Bush." In the picture, she has "Bush Cheney 04" written across her breasts. But how did conservatives get here in the first place? Why would a conservative want to act like a liberal or frequent a site dedicated to college student debauchery? This is the kind of website that conservatives should avoid as a matter of principle. The refusal to portray one's face may allow these women to deal with the cognitive dissonance that comes with espousing a conservative ideology while participating in one that is liberal. Perhaps this also points to a fracture in the conservative movement as some youth move toward a more sexually liberated ideology.

In some of the images that display the woman's face, most of the body is also visible. One Kerry supporter is shown in side profile, covering her breasts with her hands, standing next to a Kerry cutout. She has "Kerry" written on her side. The caption above her photo reads, "My titties and I are planning on voting for Mr. Kerry." A photo of another Kerry supporter takes in most of her body as well. She wears an open black sheer robe, string bikini, and black thigh-high stockings. Behind her is a large sign that says "No W." She has Kerry '04 written on her abdomen and holds a bumper sticker that states, "Be a smart ass, Vote Democrat." This fits with her caption, "Don't be a dumbass... Vote Kerry/Edwards '04." Only one image seems to have legitimate male participants, with three males and two females. The caption reads, "If you know what is right for you! you will join the revolution, and FUCK THE BUSH!" The males each have one word painted on them (fuck the bush) and the females spell out "Vote for Kerry." No one has explicit nudity—the women are covering their breasts with their hands and the males are wrapped in flags.

The strongest statement of identity and politics comes from a Kerry supporter with the caption, "My vote counts." She is standing in a shower. Her face is completely visible, as is most of her body to her upper thighs. Her shirt is pulled up to reveal her breasts. "Kerry" is written on her abdomen. She pulls her panties down slightly but nothing is revealed except a small tattoo. The statement "My vote counts" is powerful because of its simplicity and its ambiguity. Her face displays a grim expression; she appears to derive no pleasure from her exhibition. Rather, she portrays her exposure as a matter of civic duty. Her message is written far enough under her breasts that there was no need to expose them for the sake of the message. Perhaps her assertion of the validity of the vote comes from the forum in which it is expressed. Like the poll tax of yesteryear, the price of voting in this forum is feminine exposure. The caption for a photo of a Kerry supporter posted by Kyle, makes this explicit: "I would differ but i dont have tits." But this women's statement, "My vote counts," also seems to be a reaffirmation of the value of her vote. Rather than simply repeating slogans for or against the candidates, her caption emphasizes the power of the voting public. This is reinforced by the understated nature of the word on her abdomen; it is simple and unadorned with decoration. The focus is on the process rather than on the candidates. Another participant seems to take this a step further. The caption for her photo states, "Tities that voted!" In the photo, her nipples are covered with strategically placed "I voted" stickers. However, there is no indication of the candidate that she supported. Her display suggests that the act of voting is more important than the person voted for.

Prostitution and Sexual Power

The persuasive power of female sexuality is a theme that permeates many of the images, specifically the supposed convincing power of large breasts or feminine beauty. One Bush supporter has body paint with a W04 flag covering her breasts. The caption for the image reads "Bush's campaign aides." The caption for a Kerry supporter states, "Maybe I'll win over some swing voters!" She has "Vote Kerry" written on her breasts and is holding a "no W" (circle w/ slash) sign. Others simply hold the promise of more exposure if their aims are met. For example, the caption for a Kerry supporter reads "Vote kerry, and see more :)" "Kerry" is written on her exposed breasts. Another Kerry supporter writes, "If he wins, you'll see more!!!" She reveals nothing; she is wearing a bra with a Kerry pin attached. These women engage in a kind of politically motivated striptease. Others make explicit reference to the size of their breasts as a means of persuasion. A woman with "Kerry" written across her large breasts has the caption, "Does my size count for 2 votes?!" A woman with "I (heart) Bush" written across her breasts has the caption, "My fiance has two big reasons for you to vote for Bush..."

Some images contain more explicit displays of sexualized power. For example, one Bush supporter states: "Kerry suck these. Vote Bush." This seems to be a sexualized insult, but I have not previously heard anything resembling "suck my breasts" used as an insult in the same way that "suck my dick" is used by males. This seems to be a more active way of using the breasts as a site of sexual power. Another explicit image of sexual power can be seen in the image of a woman with "Lick Bush Vote Kerry" written on her torso. She is standing, wearing a bra. A person's head (which seems to be male) obscures her pubic area.

One of the more interesting instances of sexual power bears the caption: "I hope my boyfriend doesnt get mad about this...I know he looks on here but I just had to do it!" Below the caption are two grainy black and white images, one displaying the right side of her exposed torso with "Kerry" written over her nipple, and the other displaying a side view of her lower abdomen and hip with "no bush" written on her thigh. In this image, there seems to be two issues at work—the desire to participate and the right to participate. This woman feels compelled to participate and make her voice heard but expresses fear of retribution by her boyfriend. She recognizes that her boyfriend looks at the site and, presumably, at the other women on the site. The motives for her boyfriend's possible ire are purely speculative, but this woman seems to subscribe to a kind of double standard. It is acceptable for him to look at other women but it is not appropriate for her to place herself in view—there is a "bad girl/good girl" dichotomy that maintains a distinction between the other women on the site and the girlfriend. The other women are fallen, but she remains relatively pure. Such double standards are shaped early in our development and women are as likely as men to sanction other women.[27] It seems clear that this woman seems trapped between the desire to be one of these women and her perceived obligations to her boyfriend. Indeed, Shannon Gilmartin describes the confusion of college students who feel obligated to embody "the 'proper' code of feminine conduct (women are not supposed to act on their desire, especially outside of a romantic relationship)."[28] In short, the act of posting the photo illustrates a struggle over the ownership of her body. By keeping it hidden from others, her nudity remains something just for her boyfriend. Placing her naked body on display—even anonymously—is an act of defiance. The images depict very little that would be identifying. The images are grainy enough to obscure any identifying marks and her body is unremarkable in the sense that she is shaped like an average college-aged woman. She hides in a sea of flesh—the only way that she would likely be identified is if she did not remove the writing from her body or if she told him that she had posted photos of herself to the site. Even so, she makes it clear that her political actions may come at personal expense. This is so because, as Judith Butler observes, "We regularly punish those who fail to do their gender right."[29]

Some images convey explicit overtones of prostitution. Perhaps the most explicit is shown in a Kerry supporter's caption that reads, "Emily offers to buy your vote!" Below her exposed breasts is written: "Vote Kerry and I'll suck you off (girls too (Smiley face))" This is an explicit link between sexual power and political power in the form of political prostitution. Of course the system of votes in exchange for personal or group benefit is well entrenched in American politics, but generally the aim is financial gratification rather than pleasures of the flesh. A Bush supporter has the caption, "I know the quality of this pic sucks but I get $20 if its posted and plus I did the BUSH sign all by myself so make me feel special PLEASE.." Although she doesn't state who is offering her the money, there is an explicit connection between nudity, political action, and financial gain. Moreover, there is a clear desire for the affirmation of self-worth through exposure.

Three captions reveal a kind of coercion in the process, all of them for Kerry supporters. The caption for one Kerry supporter reads, "Exploiting my girlfriend for a better America. Vote Kerry." She is kneeling on a bed wearing leopard print panties and has "Vote Kerry" written on her exposed breasts. Another Kerry supporter notes, "It took me some time to get her to do it. But here's our vote for Kerry." In this caption, the male subtly claims part of the vote through the pronoun "our," thus rendering her vote not her own. Another caption reads, "Gotta love what you can get a freshman to do." This woman has a considerable amount of writing on her body. Written across her breasts and on her torso: "Collegehumor.com (heart) J.W.U. Providence." And "When all else fails, Johnson and Wales!!!" "Nernz" is written above her belly button. "Fuck Bush" is written above her pubic area and has an arrow pointing down. It is not all visible, but it seems that her pubic area is shaved. On her breast is a purple spot and "hickey" is written next to it with an arrow. This woman has become a billboard for College Humor, Johnson and Wales University, her own sexual behavior, and what may be an inside joke, as well as a way to express opinions about Bush—but whose billboard is she?

Values

A few women make explicit arguments that Bush is anti-women. One Kerry supporter has written on her torso, "Kerry A Vote for Bush = a vote against women's rights." Another Kerry supporter makes the argument more general, writing "When women vote, Democrats win" on her breasts and in her caption. Despite its prevalence in Bush's campaign ("W is for Women"), no Bush supporters used this slogan, instead using the slogans of war. For example, one woman has written on her torso, "Vote Bush if you vote for Kerry it's for a better Vietnam." Some Kerry supporters make arguments against Bush based on his handling of the "war on terror." The caption of one Kerry supporter states, "Bush once said . . . 'The world is more peaceful and more free under my leadership.' Yah, so . . . VOTE KERRY!!!!"

Some photos comment on the split between liberal and conservative and tie this to youth. For example, a photo of two Kerry supporters bears the caption, "Crazy freshman and their crazy liberal values...to bad I'm a Bush supporter. (thought I'd take one for the team and let em vote Kerry.)" It is not clear how the author of the caption is "letting them vote for Kerry," but the author paints liberal ideology as a foible of youth. Other women, especially those supporting the Libertarian candidates, are much more aggressive in expressing liberal viewpoints. One woman holds a bumper sticker beneath her breasts that states "I'm pro-choice on everything. Join the Libertarian Party Call 800-682-1776" Another Libertarian supporter caption states "BADNARIK '04!! Legalize Free Enterprise and Legalize NAKED, INTOXICATED FUN!!" Written on her torso: "Badnarik + Campagna No More Drug War."

There is dissonance in some of the photos concerning the Republican Party's preoccupation with the sanctity of marriage. In two of the photos, the women seem to be wearing wedding bands. One photo has the caption "bush is happy in the mountains." The woman has "Bush" written across her breasts and she holds a Bush doll between her breasts. She seems to have a wedding band on her left hand. Another Bush supporter holds up her tank top to expose her breasts, which have "Vote 4 Bush" written between them. Not only does she appear to have a wedding band on her left hand, she is also prominently displaying the gesture of "the shocker."[30] This act seems to encapsulate everything that Bush and the Republican Party would oppose. In another image, the caption reads, "My fiance's boobs vote for Bush," and "Vote for Bush" is written on her breasts.

The most interesting image regarding the nature of relationships can be found in a photograph of three Bush supporters. In the image, one woman has photoshopped stars covering her nipples, there are stickers on the nipples of another, and the third woman has her back to the camera. Unlike many of the images in this contest, there are subtle homoerotic undertones. The two topless women each have a hand on the buttocks of the woman in the middle. Perhaps this is an indication of the seemingly malleable sexuality of young women, but to support a party that would reject them for immoral behavior seems a peculiar strategy. However, political life is full of contradictions, as can be witnessed in such groups as the Log Cabin Republicans, a group of openly homosexual Republicans.[31]

Other Candidates

There is a sense that some Kerry supporters are not voting for him so much as voting against Bush. Several women express the sentiment "anyone but Bush." One woman has "Kerry" written on her breasts but the caption reads, "My submission for president. Pretty much anyone but Bush will do," and another wrote "Kerry because he's not Bush" on her breasts as well as in the caption. Another woman uses a sexual play on words to make her point. Her caption states, "Here's my vote for Kerry! Anything's better than Bush!" Across her breasts is written "Abstinence in 2004 No Dick + Bush." One woman expresses ambivalence over the complete election process with a commonly expressed sentiment in American politics; on her torso is written, "Kerry—the lesser of two evils."

A few women break out of the two party structure and endorse third party candidates. Four women cast their bodily vote for Nader. One woman has "Nipples for Nader" written across her breasts and has an American flag drawn on her belly. Another Nader supporter is more forceful; her caption declares, "FUCK BUSH AND THE RICH SCENE, AND I'M SURE AS HELL A KERRY HATER, GET WITH THE ORANGE, MAROON, AND GREEN . . . GET OFF YOUR ASS AND VOTE RALPH NADER!" In the photo, her nipples and pubic area are covered with photoshopped marijuana leaves. The writing on her navel states, "Down with Kerry!" and "Fuck Bush!" VT is written on her breasts and "Vote Nader" is written below. However, another woman recognizes the perils of voting for a third party candidate in what would likely be a close election. On her right breast is written, "(heart) Nader" and on the left breast, "Vote Kerry." A few women that that have already been mentioned advocated voting for the Libertarian Party candidates.

Two women take the opportunity to ignore the major candidates altogether. One woman has "Jib Jab 4 Prez!" written on her breasts and torso, referencing a popular parody video sung to the tune of Woody Guthrie's "This Land is Your Land." Making further reference to the video, she has written beneath her navel, "This land is your land" with an arrow pointing to her pubic area and her unzipped jeans that reveal her panties. In another photo, the caption declares, "Bush . . . Kerry . . . Screw Democracy, Communism all the way!!!" The woman wears a Soviet flag around her waist and has "Soviet Chick" written across her breasts, with the "e" in Soviet written backwards to create a kind of exoticness. However, in the Russian alphabet, there is no letter that is written in this manner. It seems that for both of these women, this "election" was a joke and they treated it as such by casting votes for absurdity.

The Female Body as Political Medium

Hélène Cixous writes, "Listen to a woman speak at a public gathering (if she hasn't painfully lost her wind). She doesn't 'speak,' she throws her trembling body forward; she lets go of herself, she flies; all of her passes into her voice, and it's with her body that she vitally supports the 'logic' of her speech. Her flesh speaks true. She lays herself bare. In fact, she physically materializes what she's thinking; she signifies it with her body."[32] The women in this contest quite literally inscribe their arguments on their flesh, using the body as both argument and means of argument. Although the contest may seem frivolous—a way to provide some titillation and opportunities for exhibition—it demonstrates the slippery nature of political speech. The question of whether this is simply exhibitionism masquerading as political speech is irrelevant; these images are being portrayed as political speech and, more importantly, they *are* political speech. Sonja Foss suggests that for an image to qualify as visual rhetoric, "the image must be symbolic, involve human intervention, and be presented to an audience for the purpose of communicating with an audience."[33] The College Humor images fulfill these criteria and can thus be considered visual rhetoric. Such displays, regardless of how juvenile they may appear on the surface, perform serious rhetorical work and should not be dismissed as pornography or mere exhibition.

The women in the images are simultaneously embodied and disembodied through the digital media. As Barthes observes of photography, "The photograph is literally an emanation of the referent. From a real body, which was there, proceed radiations which ultimately touch me, who am here; the duration of the transmission is insignificant; the photograph of the missing being, as Sontag says, will touch me like the delayed rays of a star. A sort of umbilical cord links the body of the photographed thing to my gaze: light, though impalpable, is here a carnal medium, a skin I share with anyone who has been photographed."[34] The viewer is invited to gaze upon each woman's naked flesh, yet what is left in the photograph is only a representation. But this does not remove the woman from her naked embodiment; Vivian Sobchack explains that the body cannot simply be relegated to the hyperreal.[35] Human experience is lived in a corporeal state, an interaction and intertwining of the real and the hyperreal, the natural and the cultural. But Monica Whitty rightly argues that "although the physical bodies are not present in cyberspace, the body still does matter."[36] Digital presence is linked to the body and the body influences one's digital presence. Thus, despite its mediated presence, the body is not reducible to its icon.

There remains an uneasy tension between the embodied actor and the mediated presentation of self. The medium allows one to adopt an online persona quite different from the one performed in his or her everyday life. A conservative Republican can behave like an exhibitionist in this space without a connection to her identity. Douglas Thomas notes that in order to try someone for a crime, or in this case to sanction someone for transgressing the moral standards of her political party, a mediated presence is not enough; "what is always needed is a body, a real body, a live body."[37]

The presentation of self is not all that is constructed in this online contest. Poststructuralist theory argues that gender roles and sexualities are socially constructed, arguing against an essentialist view of gender and sexuality.[38] This has been demonstrated through historical accounts of different gender roles in different cultures.[39] Social norms are increasingly constructed through mediated channels and this mediation affects how these norms form and shape identity.[40] Nicholas Negroponte argues that individuals are changing, becoming digital: "It is here. It is now. It is almost genetic in its nature, in that each generation will become more digital than the preceding one."[41] One reason for this is because cyberspace allows for a malleability of identity that is hard to perform in the physical world. Fortunati argues,

> In post-modern society, the social system of differences developed in the modern age is being completely restructured. Many differences, even between men and women, or more specifically, between the world of production and reproduction, have disappeared, or are at least less clear cut. There is a tendency at the social level to fusion, to the formation of hybrids, to the development of similarity. Many of these differences are artificial constructions, the result of historical, social, and cultural determinations.[42]

However, one cannot merely separate an individual's constructed digital persona and his or her embodied existence, or define one as real and the other as somehow less than real.[43]

All of this points to the interplay between technology and our conceptions of reality. How these women chose to portray themselves may signal a shift in the conception of female sexuality and political action that moves away from the idea of a passive female sexuality—her "to-be-looked-at-ness"—described by Laura Mulvey and others who have latched onto the concept of the "male gaze."[44] Even if it is, this change is not entirely unproblematic. Young notes that there has always been a wide range of responses to the gaze: "She may enjoy the attention and learn to draw the gaze to her bosom with a sense of sexual power. She may loathe and fear the gaze that fixes her in shock or mockery . . . . She may for the most part ignore the objectifying gaze, retaining nevertheless edges of ambiguity and uncertainty about her body."[45] This is borne out in the images; some of the women revel in their exhibition and others seem to beg for validation through their objectification and display. Some women externally seek for esteem, such as the woman who asked if her breasts were okay and the one who had been previously rejected by College Humor. So there still exists the tacit understanding that they are to be put on display, but in this case they are to be put on display for a reason that transcends prurient interests—because they have literally inscribed their politics upon their bodies.

But why would these women use this medium and why this forum? The simplest answer to the latter is that College Humor asked them to. Those who would be comfortable exposing themselves for a political cause had been called to action and they acted. College Humor had found that there were already women ready to expose their bodies for much less noble causes. Thus, the site had provided a way of legitimating exhibitionism in the guise of political speech. The answer to the question of why these women would use this medium is a bit more complicated. One reason that this medium would appear attractive is the commonplace nature of the naked body on the Internet. In the online world, nudity is rather easy to find. Thus, one can hide in plain sight. If one seeks female nudity for its own sake, there are many places it can be had in much more explicit forms. These images are remarkable for their non-sexualized nature. There are no photos of exposed genitalia—only breasts and pubic regions. Also, each of them is necessarily marred by the names of the candidates and the slogans that had been inscribed upon their bodies, generally in black marker. The political speech inscribed upon the naked flesh is inescapable; one is drawn to their politics rather than to their sexuality.

There is also a perception of online anonymity that makes this medium inviting. Most of these women did not display their faces, thus concealing their identity for all but those most intimate to them. Events that would actually be more fleeting—such as flashing one's breasts at a rally—may be seen, paradoxically, as more exposing. When one exposes him or herself in person, there is a presence and recognition of that person as a *particular* person. Thus, despite the short-lived nature of such an exposure, the memory of the exposure will live on in those around that person and always be connected to him or her. Disconnecting their bodies from their identities allows these women to engage in political nudity with little risk of official sanctions. In the physical world, there is the potential of being arrested for indecent exposure but in the online world amateur nudity is commonplace. They are able to project private nudity into the public sphere in a way that conceals their identity and protects them from legal action.

However, this security may come with a price. Jones argues, "Despite the boundary testing and incremental change of social norms and physical representation facilitated by mediated exhibitionism, the systemic threat to social control is mitigated by commercial concerns as authentic performances are co-opted or imitated for the sake of profit."[46] This is certainly the case for College Humor, which serves as gatekeeper for the images that appear on its site. Several of the women in the contest revealed that they had been previously rejected. By presenting these women as spectacle, College Humor may be minimizing the message itself. Much as the lactivists found in the case of Facebook, these women were required to play by the rules set out by the moderators of the site in order to have their messages disseminated. One cannot overlook the political economy of the online world.

Another aspect of the internet that makes it inviting for transgressive displays is its ephemeral nature. The College Humor contest has likely been forgotten by now, replaced by still more contests asking women to bare their bodies for any number of reasons. Although the official page for the contest has been taken down, the images are still available in the archives of the site. The perceived lack of permanence in online communication may provide an opportunity to step outside of the norms with the idea that no one is really looking anyway and it will soon be lost and forgotten. But Fernback notes that "Cyberspace is a repository for collective cultural memory—it is popular culture, it is narratives created by its inhabitants that remind us who we are, it is life as lived and reproduced in pixels and virtual texts."[47] Whether they realize it or not, these women are participating in the battle over the construction of sexual norms. By exposing themselves, they add to the flood of images that legitimizes nudity, rejecting what John Ince calls "fig-leafing."[48] As such, they are taking an active role in the display of their body. They are not simply being looked at; they *place* themselves in the line of sight of others. Thus they work to reconfigure the sexual norms of femininity away from passive objectification and toward a more active role, claiming their body and using it to augment and amplify their political views. They break the taboo of public nudity in order to demonstrate the strength of their commitment. This may seem to be a romantic view of these images, but many of women portray the act more as political expression than mere exhibition. Because of the internet's role as a repository of public memory, these images will always be political in that they help to reconfigure views of female nudity.

Perhaps the most important aspect of the medium is the ability to control the exposure. Foss argues that in visual rhetoric "visual elements are arranged and modified by a rhetor not simply for the purposes of emotional discharge but for communication with an audience, even if the creator is the sole audience for the image or object."[49] In the online world, one can construct the persona, the magnitude of exposure, and even digitally alter the image to "correct" skin tone, remove blemishes, and crop to hide revealing characteristics. The woman exposes herself, but only what she wishes to expose. Moreover, she can expose herself to many in the digital world while exposing herself to no one in the physical world. In short, she gains the benefits of exposure while minimizing the risks. As Yvonne Jewkes and Keith Sharp observe, the Internet "can liberate its users from the usual constraints of corporality. The Internet thus gives users a freedom of expression—a freedom of *being*—quite unlike anything they have at their disposal in the physical world."[50] By anonymously displaying one's body online, these women are able to (somewhat) safely transgress taboos against nudity and exhibitionism to make a political statement.

In these images, the body is displayed but mainly as a means of disseminating a message. The display is not connected to the individual in question; few individuals can be identified by their breasts alone. Thus, this display is an anonymous display which reduces the icon of the woman to a symbol of womanhood. Richard Sennett connects citizenship with the idea of civility, defining civility as "the activity which protects people from each other and yet allows them to enjoy each other's company. Wearing a mask is the essence of civility."[51] Adopting an anonymous persona allows for the kind of impersonal interaction championed by Sennett. To mask or hide the face, and thus the identity of the woman, allows her to display the message with only the body to enhance or detract from the message.

And the use of the body may very well enhance the message. John Poulakos argues that "the Sophists conceived of rhetoric primarily as a *technē* (art) whose medium is *logos* and whose double aim is *terpsis* (aesthetic pleasure) and *pistis* (belief)."[52] Charles Segal explains how aesthetic pleasure amplifies the persuasive power of speech: "There is rather a psychic complicity in the emotive action of the *logos*: the psyche participates in and reacts to the artistic composition of the *logos* and thus experiences *terpsis*; it is hence regarded as a perceptive, aesthetically sensitive organ upon which the work of art acts. When the aesthetic stimulus is strong enough, however, as in the case of a pleasing vision or a moving speech, the passive aesthetic *terpsis* becomes a powerful impulse which directs the whole course of action of the psyche."[53] Logic combined with beauty can be a potent rhetorical strategy.

Many of the participants in the College Humor contest seemed convinced that their attractiveness could prove persuasive to others. They may be on to something; Robert Cialdini and Noah Goldstein suggest that "one of the clearest implications of our desire to affiliate with others is that the more we like and approve of them, the more likely we are to take actions to cultivate close relationships with them."[54] Indeed, some research suggests that people are more likely to comply with requests when they are made by attractive individuals than by those who are unattractive.[55] But attractiveness is not always an asset. Although Stefanie Johnson and colleagues found that "attractiveness is beneficial for men and women applying for most jobs," they also observed that "attractiveness could be detrimental for attractive women applying for masculine jobs for which physical appearance was perceived to be unimportant."[56]

It is difficult to determine exactly how persuasive these strategies would be in actually inducing a change in others' opinions concerning which candidate is more desirable. In some ways, these bodily displays function in much the same way that lawn signs for particular candidates do—as social proof. Robert Cialdini explains that "one means we use to determine what is correct is to find out what other people think is correct."[57] In other words, the more people that hold a particular stance, the more likely it is that others will come to view that stance as desirable or correct. In this case, if more people side with a particular candidate, that candidate is likely to be seen as desirable. It is easy to see the Libertarian and independent candidates as unelectable, despite the fact that some of these participants may actually share some of the tenets of their political philosophy, because so few participants provided their voice (or bodies) in support of these candidates. But the social proof perspective, when taken together with the focus on attraction, may be counterproductive. Drawing on an evolutionary model, Vladas Griskevicius and colleagues found that "romantic desire caused normally effective social proof appeals to backfire, though the same social proof appeals were more effective following fear-inducing content."[58]

Studies examining how nudity functions as a means of persuasion also suggest that this strategy may yield mixed results. Jessica Severn, George Belch, and Michael Belch found that when using sexual appeals brand recall was not affected, but recall of copy was hindered.[59] Thus, in this contest, those who had only the candidate's name would likely find that their nudity would not hinder their message, but those who provided more lengthy textual arguments may find their message overpowered by the display of their body. Moreover, research suggests that whether a message is viewed favorably partially depends on whether the nudity is congruent with the message.[60] Unlike cosmetics or high end clothing brands that promise to increase sex appeal, nudity has little to do with politics, so one could expect that the strategy would be viewed as gratuitous. Nudity is, however, somewhat common on College Humor, which troubles the notion of what is really being promoted in these images—especially in light of the many inscribed statements expressing love for College Humor or the woman's school. This also points to the importance of considering the implied reader of College Humor; Jaideep Sengupta and Darren Dahl found that liberal attitudes toward sex—which seems consistent with the ideology put forth within College Humor—contributed to increased liking of advertisements featuring gratuitous sexual appeals.[61] At any rate, the conventional wisdom that sex sells remains in force in many of these images, with women expressing the belief that the quality of their bodies could persuade others to share their political views.

Finally, these images may fail as persuasive tools because of the constraints of the media—both the online environment and the body itself. With limited space in the caption, the only place left to argue is the body itself. The body, however, can only hold so much information, especially when it is being photographed. Thus it is little wonder that the women are reduced to relying on ad hominem attacks and slogans rather than on actual arguments that differentiate the policies and platform of their chosen candidate from the others in the field. Much like a billboard, the message must capture the attention of the viewer but cannot rely too heavily on text. Yet the differences between the candidates warrant scrutiny, and these kinds of truncated arguments do little to actually enrich the public sphere. Seyla Benhabib argues that "the autonomous citizen, whose reasoned judgment and participation was the sine qua non of the public sphere, has been transformed into the 'citizen consumer' of packaged images and messages or the 'electronic mail target' of large lobbying groups and organizations."[62] Reducing the complexity of political discourse allows citizens to see issues as either/or propositions and different viewpoints as representing a *them* opposed to an *us*. Such body rhetoric does little to interrogate the differences among the candidates, and instead simply marks each woman as a member of a particular faction. Like the ballots that these women will presumably cast, there is little required of them but to make a choice between the candidates.

It seems reasonable that the exposed body will function differently in a mediated context, especially as these media influence societal norms. Jean Baudrillard argues that "we are witnessing the end of perspectival and panoptic space . . . and thus to the *very abolition of the spectacular.*"[63] But perhaps it is less an end to panoptic space and more an indifference concerning our surveillance. In these images, we see the body willingly put on display. We invite and demand the gaze. Susan Bordo describes the "receptive pleasures traditionally reserved for women," such as "the pleasures, not of staring someone down but of feeling one's body caressed by another's eyes . . . Some people describe these receptive pleasures as 'passive' . . . 'Passive' hardly describes what's going on when one person offers himself or herself to another. Inviting, receiving, responding—these are active behaviors too, and rather thrilling ones."[64]

But these women seem to be engaging in something other than simple exhibitionism and the fulfillment of the desire for the limelight of the gaze. Rather, this seems to be a way to express one's political views and attempt to persuade others through the non-traditional means of imagery. J. Anthony Blair notes that "the advantage of visual arguments over print or spoken arguments lies in their evocative power," and suggests that one aspect of the power of visual imagery lies in its ability to "evoke involuntary reactions—reactions that must be counteracted by the recipient if their power is to be at all diffused."[65] Women recognize that they wield sexual power and body display is one way to incite desire in others. This alone is not problematic, but if women feel *required* to use their sexuality in order to have their political voices heard, then there is something wrong with the means by which the citizenry—especially the female half—is heard.

Despite its potential to transcend the constraints of corporeality, mediation places other constraints on the individual. Also, one cannot completely transcend embodiment; the mediated body still exists in the physical world. This form of political speech forces one to recognize that those who take part in political processes are individuals and not simply numbers or aggregates. Each woman here is an individual with her own beliefs, thoughts, ideas, feelings, and physical embodiment. When she removes her clothing, she is just her—a woman. This contest reminds the viewer, perhaps unintentionally, that political processes affect human beings—vulnerable, naked, *real* human beings.

Notes

1. The contest has since vanished into the dark corners of the website's archives, but many of the images can still be found on the site. I archived the images while the photos from the contest were still easily available. I have kept all of the captions as originally on the site. Although there are many misspellings, I have avoided the use of [*sic*] to maintain ease of reading.

2. John Pratt, "Pornography and Everyday Life," *Theory, Culture & Society* 3, no. 1 (1986): 72.

3. Ruth Barcan, "Home on the Rage: Nudity, Celebrity, and Ordinariness in the Home Girls/Blokes Pages," *Continuum: Journal of Media & Cultural Studies* 14, no. 2 (2000): 146.

4. This is not to say that liberal movements have a monopoly on such events. There have been some protest actions that take a stance that would seem out of place among liberals, such as a nude display in favor of genetically modified food. See "With Friends

Like These, Monsanto Needs No Enemies," USA Today, www.usatoday.com/tech/news /techpolicy/2003-08-06-raelians-biotech_x.htm. See also Nick O'Malley and Jonathan Pearlman, "Raelian around the GM Flag," *Sydney Morning Herald* August 8, 2003, 16.

5. See Kevin Michael DeLuca, "Unruly Arguments: The Body Rhetoric of Earth First!, ACT UP, and Queer Nation," *Argumentation & Advocacy* 36, no. 1 (1999): 9-21.

6. Roland Barthes, *Camera Lucida: Reflections on Photography*, trans. Richard Howard (New York: Hill and Wang, 1981), 34.

7. Robert Asen, "A Discourse Theory of Citizenship," *Quarterly Journal of Speech* 90, no. 2 (2004): 191.

8. Gerard A. Hauser, "Incongruous Bodies: Arguments for Personal Sufficiency and Public Insufficiency," *Argumentation & Advocacy* 36, no. 1 (1999): 2.

9. Roland Barthes, *Image, Music, Text*, trans. Stephen Heath (New York: Hill and Wang, 1977), 38-39.

10. See Shannon K. Gilmartin, "Changes in College Women's Attitudes toward Sexual Intimacy," *Journal of Research on Adolescence* 16, no. 3 (2006): 429-54; Brett Lunceford, "The Walk of Shame: A Normative Description," *ETC: A Review of General Semantics* 65, no. 4 (2008): 319-29; Brett Lunceford, "Smeared Makeup and Stiletto Heels: Clothing, Sexuality, and the Walk of Shame," in *College Sex: Philosophy for Everyone: Philosophers with Benefits*, ed. Robert Stewart and Michael Bruce, 51-60 (Hoboken, NJ: Wiley-Blackwell, 2010).

11. Panteá Farvid and Virginia Braun, "'Most of Us Guys Are Raring to Go Anytime, Anyplace, Anywhere': Male and Female Sexuality in *Cleo* and *Cosmo*," *Sex Roles* 55, no. 5 (2006): 307.

12. Bahar Hanjani, "The Illustration of Beauty: Super-Exposed in the U.S., Veiled in Iran," *Semiotica* 187, no. 1-4 (2011): 310.

13. Hanjani, "The Illustration of Beauty," 320.

14. For more on how these images are perceived among cultures, see Amir Hetsroni, "Sexual Content on Mainstream TV Advertising: A Cross-Cultural Comparison," *Sex Roles* 57, no. 3/4 (2007): 201-10; Michelle R. Nelson and Hye-Jin Paek, "Nudity of Female and Male Models in Primetime TV Advertising across Seven Countries," *International Journal of Advertising* 27, no. 5 (2008): 715-44; Hye-Jin Paek and Michelle R. Nelson, "A Cross-Cultural and Cross-Media Comparison of Female Nudity in Advertising," *Journal of Promotion Management* 13, no. 1/2 (2007): 145-67. For more on nudity and advertising in general, see Andrea C. Beetles and Lloyd C. Harris, "Consumer Attitudes Towards Female Nudity in Advertising: An Empirical Study," *Marketing Theory* 5, no. 4 (2005): 397-432; Sid C. Dudley, "Consumer Attitudes toward Nudity in Advertising," *Journal of Marketing Theory and Practice* 7, no. 4 (1999): 89-95; Michael S. LaTour, "Female Nudity in Print Advertising: An Analysis of Gender Differences in Arousal and Ad Response," *Psychology & Marketing* 7, no. 1 (1990): 65-81; Jessica Severn, George E. Belch, and Michael A. Belch, "The Effects of Sexual and Non-Sexual Advertising Appeals and Information Level on Cognitive Processing and Communication Effectiveness," *Journal of Advertising* 19, no. 1 (1990): 14-22; Claire Sherman and Pascale Quester, "The Influence of Product/Nudity Congruence on Advertising Effectiveness," *Journal of Promotion Management* 11, no. 2/3 (2005): 61-89; Lawrence C. Soley and Leonard N. Reid, "Taking It Off: Are Models in Magazine Ads Wearing Less?" *Journalism Quarterly* 65, no. 4 (1988): 960-66. There is some research also focusing on male nudity in advertisements, but Penny Simpson and colleagues concede that "despite the increased use of males portrayed in sexual and disparaging roles in advertisements, very little research has been conducted to determine effects of the ad execution on con-

sumer responses to the ads." Penny M. Simpson, Steve Horton, and Gene Brown, "Male Nudity in Advertisements: A Modified Replication and Extension of Gender and Product Effects," *Journal of the Academy of Marketing Science* 24, no. 3 (1996): 257.

15. See Lemi Baruh, "Publicized Intimacies on Reality Television: An Analysis of Voyeuristic Content and Its Contribution to the Appeal of Reality Programming," *Journal of Broadcasting & Electronic Media* 53, no. 2 (2009): 190-210; James Hibberd, "Poll: TV Violence Worse Than Nudity," *Television Week* 24, no. 4 (2005): 59.

16. Matthew T. Jones, "Mediated Exhibitionism: The Naked Body in Performance and Virtual Space," *Sexuality & Culture* 14, no. 4 (2010): 256.

17. Liam Stack and David D. Kirkpatrick, "Nude Blogger Riles Egyptians of All Stripes," *New York Times* November 18, 2011, 4.

18. Haig A. Bosmajian, "Defining the 'American Indian': A Case Study in the Language of Suppression," *Speech Teacher* 22, no. 2 (1973): 89.

19. Leopoldina Fortunati, "The Human Body: Natural and Artificial Technology," in *Machines That Become Us: The Social Context of Personal Communication Technology*, ed. James Everett Katz, 71-87 (New Brunswick, NJ: Transaction Publishers, 2003), 75.

20. Valerie V. Peterson, "The Rhetorical Criticism of Visual Elements: An Alternative to Foss's Schema," *Southern Communication Journal* 67, no. 1 (2001): 20.

21. See Barbara J. Phillips and Edward F. McQuarrie, "Beyond Visual Metaphor: A New Typology of Visual Rhetoric in Advertising," *Marketing Theory* 4 (2004): 122.

22. See Magdala Peixoto Labre, "The Brazilian Wax: New Hairlessness Norm for Women?" *Journal of Communication Inquiry* 26, no. 2 (2002): 120.

23. Charles S. Peirce, *Collected Papers*, ed. Charles Hartshorne and Paul Weiss, 6 vols. (Cambridge, MA: Belknap Press of Harvard University Press, 1960), 2:248.

24. Jones, "Mediated Exhibitionism," 256.

25. Iris Marion Young, *On Female Body Experience: "Throwing Like a Girl" and Other Essays* (New York: Oxford University Press, 2005), 77.

26. Jones, "Mediated Exhibitionism," 263.

27. See Deborah Chambers, Estella Tincknell, and Joost Van Loon, "Peer Regulation of Teenage Sexual Identities," *Gender and Education* 16, no. 3 (2004): 397-415.

28. Gilmartin, "Changes in College Women's Attitudes toward Sexual Intimacy," 429-30.

29. Judith Butler, *Gender Trouble: Feminism and the Subversion of Identity* (New York: Routledge, 1990), 140.

30. The "shocker" is a gesture that is common in photos posted to College Humor. It is formed by holding down the ring finger with the thumb. The gesture is an allusion to placing the first two fingers into a woman's vagina and the pinky into her anus, or as College Humor puts it, "two in the pink, one in the stink."

31. Chris Barron, "A Big Question for Gay Republicans," *Gay & Lesbian Review Worldwide* 11 (2004): 27-29.

32. Hélène Cixous, "The Laugh of the Medusa," in *The Rhetorical Tradition: Readings from Classical Times to the Present*, ed. Patricia Bizzell and Bruce Herzberg, 1524-36 (Boston, MA: Bedford/St. Martin's, 2001), 1528.

33. Sonja K. Foss, "Theory of Visual Rhetoric," in *Handbook of Visual Communication Research: Theory, Methods, and Media*, ed. Ken Smith, Sandra Moriarty, Keith Kenney, and Gretchen Barbatsis, 141-52 (Mahwah, NJ: Lawrence Erlbaum, 2005), 144.

34. Barthes, *Camera Lucida*, 80-81.

35. Vivian Sobchack, "Beating the Meat/Surviving the Text, or How to Get out of This Century Alive," in *Cyberspace/Cyberbodies/Cyberpunk: Cultures of Technological*

Embodiment, ed. Mike Featherstone and Roger Burrows, 205-14 (London: Sage, 1995); Vivian Sobchack, *Carnal Thoughts: Embodiment and Moving Image Culture* (Berkeley: University of California Press, 2004).

36. Monica Therese Whitty, "Cyber-Flirting: Playing at Love on the Internet," *Theory & Psychology* 13, no. 3 (2003): 345.

37. Douglas Thomas, *Hacker Culture* (Minneapolis: University of Minnesota Press, 2002), 182.

38. See Sandra Lipsitz Bem, "Dismantling Gender Polarization and Compulsory Heterosexuality: Should We Turn the Volume Down or Up?" *Journal of Sex Research* 32, no. 4 (1995): 329-34; Judith Butler, *Bodies That Matter: On the Discursive Limits of 'Sex'* (New York, NY: Routledge, 1993); Butler, *Gender Trouble*; Young, *On Female Body Experience*.

39. See Magnus Hirschfeld, *Curious Sex Customs in the Far East: The World Journey of a Sexologist*, trans. O. P. Green (New York: Capricorn Books, 1935).

40. See Marshall McLuhan, *Understanding Media: The Extensions of Man* (Cambridge, MA: MIT Press, 1994); Marshall McLuhan, Quentin Fiore, and Jerome Agel, *The Medium Is the Massage: An Inventory of Effects* (San Francisco: HardWired, 1996).

41. Nicholas Negroponte, *Being Digital* (New York: Knopf, 1995), 231.

42. Fortunati, "The Human Body," 79.

43. See Ollivier Dyens, *Metal and Flesh: The Evolution of Man: Technology Takes Over* (Cambridge, MA: MIT Press, 2001); Sherry Turkle, *Life on the Screen: Identity in the Age of the Internet* (New York: Simon & Schuster, 1995).

44. Laura Mulvey, "Visual Pleasure and Narrative Cinema," *Screen* 16, no. 3 (1975): 6-18.

45. Young, *On Female Body Experience*, 77-78.

46. Jones, "Mediated Exhibitionism," 267.

47. Jan Fernback, "The Individual within the Collective: Virtual Ideology and the Realization of Collective Principles," in *Virtual Culture: Identity and Communication in Cybersociety*, ed. Steve Jones, 36-54 (London: Sage Publications, 1997), 37.

48. John Ince, *The Politics of Lust* (Amherst, NY: Prometheus Books, 2005), 17.

49. Sonja K. Foss, "Framing the Study of Visual Rhetoric: Toward a Transformation of Rhetorical Theory," in *Defining Visual Rhetorics*, ed. Charles A. Hill and Marguerite H. Helmers, 303-13 (Mahwah, NJ: Lawrence Erlbaum, 2004), 305.

50. Yvonne Jewkes and Keith Sharp, "Crime, Deviance and the Disembodied Self: Transcending the Dangers of Corporeality," in *Dot.Cons*, ed. Yvonne Jewkes, 1-14 (Portland, OR: Willan Publishing, 2003), 3.

51. Richard Sennett, *The Fall of Public Man* (New York: W.W. Norton, 1996), 264.

52. John Poulakos, "Toward a Sophistic Definition of Rhetoric," *Philosophy & Rhetoric* 16, no. 1 (1983): 36.

53. Charles P. Segal, "Gorgias and the Psychology of the Logos," *Harvard Studies in Classical Philology* 66 (1962): 126.

54. Robert B. Cialdini and Noah J. Goldstein, "Social Influence: Compliance and Conformity," *Annual Review of Psychology* 55, no. 1 (2004): 598.

55. Shelly Chaiken, "Communicator Physical Attractiveness and Persuasion," *Journal of Personality & Social Psychology* 37, no. 8 (1979): 1387-97; Matthias Messner, Marc-André Reinhard, and Siegfried Ludwig Sporer, "Compliance through Direct Persuasive Appeals: The Moderating Role of Communicator's Attractiveness in Interpersonal Persuasion," *Social Influence* 3, no. 2 (2008): 67-83; Peter H. Reingen and Jerome B. Kernan, "Social Perception and Interpersonal Influence: Some Consequences of the

Physical Attractiveness Stereotype in a Personal Selling Setting," *Journal of Consumer Psychology* 2, no. 1 (1993): 25-38; Marc-André Reinhard, Matthias Messner, and Siegfried Ludwig Sporer, "Explicit Persuasive Intent and Its Impact on Success at Persuasion—The Determining Roles of Attractiveness and Likeableness," *Journal of Consumer Psychology* 16, no. 3 (2006): 249-59. However, some research has challenged the idea that attractiveness leads to greater persuasion. Marjorie Caballero and colleagues suggest that "positive feelings do not necessarily translate into actual behaviors or even buying intentions." Marjorie J. Caballero, James R. Lumpkin, and Charles S. Madden, "Using Physical Attractiveness as an Advertising Tool: An Empirical Test of the Attraction Phenomenon," *Journal of Advertising Research* 29, no. 4 (1989): 21.

56. Stefanie K. Johnson, Kenneth E. Podratz, Robert L. Dipboye, and Ellie Gibbons, "Physical Attractiveness Biases in Ratings of Employment Suitability: Tracking Down the 'Beauty Is Beastly' Effect," *Journal of Social Psychology* 150, no. 3 (2010): 313, 314.

57. Robert B. Cialdini, *Influence: The Psychology of Persuasion*, Rev. ed. (New York: Morrow, 1993), 116.

58. Vladas Griskevicius, Noah J. Goldstein, Chad R. Mortensen, Jill M. Sundie, Robert B. Cialdini, and Douglas T. Kenrick, "Fear and Loving in Las Vegas: Evolution, Emotion, and Persuasion," *Journal of Marketing Research* 46, no. 3 (2009): 392.

59. Jessica Severn, George E. Belch, and Michael A. Belch, "The Effects of Sexual and Non-Sexual Advertising Appeals and Information Level on Cognitive Processing and Communication Effectiveness," *Journal of Advertising* 19, no. 1 (1990): 22, 21.

60. See Claire Sherman and Pascale Quester, "The Influence of Product/Nudity Congruence on Advertising Effectiveness," *Journal of Promotion Management* 11, no. 2/3 (2005): 61-89. This was also reinforced in Reichert's overview of sex and advertising research. See Tom Reichert, "Sex in Advertising Research: A Review of Content, Effects, and Functions of Sexual Information in Consumer Advertising," *Annual Review of Sex Research* 13 (2002): 241-73.

61. Jaideep Sengupta and Darren W. Dahl, "Gender-Related Reactions to Gratuitous Sex Appeals in Advertising," *Journal of Consumer Psychology* 18, no. 1 (2008): 73.

62. Seyla Benhabib, "Models of Public Space: Hannah Arendt, the Liberal Tradition, and Jürgen Habermas," in *Habermas and the Public Sphere*, ed. Craig J. Calhoun, 73-98 (Cambridge, MA: MIT Press, 1992): 93.

63. Jean Baudrillard, *Simulacra and Simulation* (Ann Arbor: University of Michigan Press, 1994), 30.

64. Susan Bordo, *The Male Body: A New Look at Men in Public and in Private* (New York: Farrar, Straus and Giroux, 1999), 190.

65. J. Anthony Blair, "The Rhetoric of Visual Arguments," in *Defining Visual Rhetorics*, ed. Charles A. Hill and Marguerite H. Helmers, 41-61 (Mahwah, NJ: Lawrence Erlbaum, 2004), 51, 54.

Chapter 6
Conclusion: The Nude Body in the Public Sphere

Debra Hawhee writes, "Rhetoric's very tendency to stretch or spill over into other arts or disciplines suggests what nearly everyone already knows—almost intuitively—that rhetoric isn't just a cerebral, conscious process, that it's messy, unpredictable, and that, at some level at least, the body is involved."[1] Although there is a growing interest in how the body functions rhetorically, there remains an impulse among rhetorical scholars to collapse the body into a particular textual formation. This can, in some ways, be attributed to the long history of privileging *logos* over *pathos*, the rational over the irrational—a sentiment that reaches back to Aristotle's *Rhetoric*. Bodies, however, are not so easily dismissed and they do not behave in the same manner as texts or images. In order to fully account for the rhetoric of the body, one must examine the body not only as it is constructed through rhetoric, but also how individuals use the body to construct rhetoric itself.

This book has examined how people use their bodies for political and social aims, whether intentionally or unintentionally. Those such as PETA, lactivists, and the World Naked Bike Ride employ the use of the body strategically and intentionally. The College Humor contributors engage in a form of unconventional political action that not only seeks to proclaim their views on the presidential race, but also calls into question the traditional view of female passivity. What all of these groups share is the notion that something can happen when you take your clothes off and that the act of disrobing can have social and political consequences. This makes disrobing in these contexts a rhetorical move, and Parke Burgess suggests that "the strategies and motives of any rhetoric . . . represent an invitation to a life-style, an invitation to adopt a pattern of strategies and motives, verbal and nonverbal, that determine how men and women will function together in culture."[2] Indeed, Edwin Black notes that rhetorical discourse features "enticements not simply to believe something, but to *be* something."[3]

First we must begin by considering how the unclothed body functions in a political sense. Considering that everyone has a body under his or her clothes, one must begin with the obvious question: what's the big deal? In some cultures bare breasts are hardly even acknowledged, yet in others a woman wearing a t-shirt would be unthinkable because such a garment exposes too much. The role of culture is inescapable. In American society there is a long tradition of simultaneously celebrating and denigrating the body. Moreover, there is a culturally formed connection between nudity and sexuality; sexuality, in turn, has been separated from the rest of human experience, something that sex researchers William Masters and Virginia Johnson call an "unfortunate though understandable by-product of the cultural residual of Puritan ethic (or any other 'sex is sin' concept)."[4] In answer to her own question, "Why are so many spiritual movements obsessed with eradicating the sexual?" sex writer Susie Bright responds, "Repression of the body is prerequisite to subordination."[5] Those who most vehemently oppose nude protest express concerns over the power of the exposed body to affect others in dangerous ways. In short, the body is suspect and has been since Adam and Eve partook of the forbidden fruit; thus "we distrust the body, and so we constantly watch it as if it were something separate from us."[6]

From a theological perspective, the magnitude of sinfulness that is generally ascribed to nudity is somewhat suspect. In the second book of Genesis, we read the account of the creation of Adam and Eve, which concludes, "And they were both naked, the man and his wife, and were not ashamed."[7] God created man and woman in a state of nakedness. It was not until Adam and Eve partook of the tree of knowledge of good and evil that they saw nakedness as something of which to be ashamed. In the account found in the third book of Genesis, we read:

> And the eyes of them both were opened, and they knew that they were naked; and they sewed fig leaves together, and made themselves aprons. And they heard the voice of the LORD God walking in the garden in the cool of the day: and Adam and his wife hid themselves from the presence of the LORD God amongst the trees of the garden. And the LORD God called unto Adam, and said unto him, Where art thou? And he said, I heard thy voice in the garden, and I was afraid, because I was naked; and I hid myself. And he said, Who told thee that thou wast naked? Hast thou eaten of the tree, whereof I commanded thee that thou shouldest not eat?[8]

It seemed that the nudity itself was less of a concern than their disobedience. In knowing that they were naked, Adam and Eve demonstrated their sudden lack of innocence by covering themselves. Later on we read that, "Unto Adam also and to his wife did the Lord God make coats of skins, and clothed them."[9] However, on driving Adam and Eve out of the garden we read, "And the Lord God said, Behold, the man is become as one of us, to know good and evil."[10] It is not a large theological leap to ask why, if God knew good and evil, he did not also recognize nudity as problematic and clothe Adam and Eve upon creating them. Some theologians question the idea that nudity itself was the problem. One argues, "It was not nakedness of the body that troubled them, but nakedness of the soul."[11] Another asks, "Are we not perfectly acceptable in our original nakedness? Is not our original condition, as made by God, good enough for us? In the end it is our inability to cope with the threat—for such it is—of our own nakedness which is the root of tragedy and disaster in human experience. . . . It is this inability which drives us out of the Garden of Eden into a world of our own making."[12] Yet another theologian suggests that genitals became shameful when they became "private."[13] Some groups, such as the Garden Brethren, "literally adopted nudity in their quest for innocence and authenticity"[14] and the Adamites, who claimed that nudity was "a true 'Garden of Eden' condition (when humans' relationship with God was an ideal one) and should therefore be emulated."[15] Even so, the notion that humanity had become inherently depraved subsequent to the fall of Adam and Eve remains firmly entrenched and the flesh is a core component of that depravity.[16] The story of Adam and Eve becomes one of childlike innocence forever lost; one cannot go back to the garden.

The notion that the body can lead one to sin can sometimes be taken to surprising extremes. Masters and Johnson relate the case of a young woman who was raised in a strictly religious home and was "taught that almost any physical expression might be suspect of objectionable sexual connotation. For example, she was prohibited when bathing from looking at her own breasts either directly or from reflection in the mirror for fear that unhealthy sexual thoughts might be stimulated by visual examination of her own body."[17] Thus it is not only looking at another body, but even one's own that can lead to lustful thoughts, and this lust—or potential for lust—seems at the heart of the issue of how the body influences public deliberation. Saint Augustine writes, "Such lust does not merely invade the whole body and outward members; it takes such complete and passionate possession of the whole man, both physically and emotionally, that what results is the keenest of all pleasures on the level of sensation; and, at the crisis of excitement, it practically paralyzes all power of deliberate thought."[18]

This incapacitation of deliberate thought in the face of lust is a key element of how the unclothed body functions in the public sphere. Despite Jürgen Habermas's assertion that "moral questions can in principle be decided rationally, i.e., in terms of *justice* or the generalizability of interests," the act of disrobing makes rational discussion much more difficult because there are now three moral questions—the one that sparked the protest, the question of the morality of nudity, and the morality of the observer.[19] In other words, the morality of all parties involved is called into question even as one deliberates concerning the morality of a particular course of action. Can one deliberate rationally when overcome with emotion?

In Aristotle's discussion of rhetoric, he observed that there were three core bases of persuasion: *ethos*, *pathos*, and *logos*. *Logos* is the argument itself, the words, the logic. Concerning *pathos*, or emotional appeals, Aristotle dryly observes, "Our judgements when we are pleased and friendly are not the same as when we are pained and hostile."[20] In other words, the emotional response induced by the display will influence the viewers' judgment concerning the message. As such, *logos* is not so easily separated from *pathos*. As Mitchell Berbrier writes, "Logos and pathos are distinct only in philosophical space; in the social world, intersubjective agreement is always achieved patho-logically, in cultural space and affective context."[21] Still, the elicited emotion is not the same as the argument made, and in this sense the emotional element is, to an extent, out of the control of both the rhetor and the audience. In the case of the PETA protests, for example, the activists seem intent on inducing feelings of pity for the captive circus animals, but the message was overpowered by feelings of lust for the nude protestor. Once the demonstration is released into the public sphere, it is no longer in control of the rhetor.[22] But some of these emotional elements of rhetorical discourse seem to be out of the control of the audience as well. The body's fight or flight reflex can spring into action well before we even know the nature of the danger, and it seems clear that people view the nude body in public with a sense of shock.[23]

In addition to the argument itself and emotional appeals, the third basis of persuasion is *ethos*, or persuasion generated by the rhetor. Aristotle defines *ethos* thus: "Persuasion is achieved by the speaker's personal character when the speech is so spoken as to make us think him credible. . . . This kind of persuasion, like the others, should be achieved by what the speaker says, not by what people think of his character before he begins to speak."[24] In other words, *ethos* is created in the act of deliberation. When the individual attempting to persuade disrobes, they behave in such a way as to alter the listeners' conception of who the rhetor is. Action is communicative; Gary Fine notes that "talk, while powerful, is generally perceived to be somewhat less effective in cementing loyalty and a sense of community than is physical action: putting words into deeds in a public domain."[25] What, then, is the viewer invited to believe concerning the rhetor in the case of disrobing? To put it another way, when the voice is silent the body still speaks—what does the body say when it is denuded? It seems that there are four statements that answer this question: look at me; desire me; fear me; believe me. I will consider each of these in turn.

Look At Me

It seems obvious that those disrobing seek the attention that nudity provides. Some of the individuals and organizations examined in this book explicitly state this motive. There is a pitfall to this statement, however, in that one may be seen as less than sincere—merely an "attention whore." It is easy to dismiss the disrobing as not a signal that draws others to their message, but rather a strategy to cast the limelight on him or herself. The equivocality of the body allows for multiple interpretations.

When an individual exhibits otherwise hidden flesh, he or she issues a demand that others look. Indeed, when confronted with a naked body, the natural instinct would be to look, partly out of a sense of shock. It is quite uncommon to find a nude individual in public, so the sense of novelty functions as a way to draw in the viewer and capture attention. Moreover, humans seem to possess an innate desire to look at people, especially at things that they are not supposed to see.

Despite the common notion that exhibitionists simply expose themselves because they are sexual deviants, some research has called this notion into question.[26] However, whether or not this is empirically verifiable is of less concern than what the body is communicating. Because of the strong taboo on exposure of those parts of the body coded as sexual, the command to "look at me" can really be translated to "look at me express my sexuality," or simply "look at my sex organs." Those who engage in these acts likely do so because of the certainty that it will be noticed. They cannot help but see that even those who have faded from the limelight can have its warm glow cast upon them again simply by providing a bit of exposure. But there are many ways to compel others to look without resorting to disrobing. Choosing the strategy of nudity calls into question one's innocence.

We cannot lose sight of the fact that in many of the cases examined in this book, the nudity is an intrusion. How people will react to such intrusion depends on a host of variables, including religious beliefs, comfort with nudity, and gender. One study examined online intrusions of sexually explicit material and found that females "held less positive attitudes toward the unsolicited sexually explicit material for all three online sources, inbox messages, junk mail messages, and pop-ups, when compared to their male counterparts."[27] Nudity and sexuality are, in fact, different but they are often conflated in Western culture, thus this may provide clues concerning the effectiveness of this particular form of shock activism. If females are more likely to view unexpected nudity in an unfavorable way, this may account for some feminist critiques of PETA's use of nude protest.[28]

This imperative to have others look may have implications not only for those looking, but also for how one views the self. As Ann Fischer and her colleagues note, "A woman may adapt by developing an observer's perspective on her own body, engaging in habitual self-surveillance, and over time, involuntarily devoting cognitive and affective resources to the recurring implicit question 'How do I look from the outside?' Attention shifts from what she is seeing as an active subject and toward how she is being seen."[29] This seems especially likely in the case of nude displays of the self, in which one is more likely to be painfully aware of the inability to conceal self-perceived body flaws. As such, when placing oneself on display, he or she may lose sight of what they wish to convey and become more concerned with how others may view the display. Although this may be a valid rhetorical strategy if focused on the message being conveyed, this becomes problematic when the individual is focused more on his or her own appearance. When this happens, the nudity no longer becomes a means to an end, but the display becomes the desired end itself. In essence, the action one attempts to induce in the audience is no longer "look at me so you will understand my message," but rather "look at *me*."

Desire Me

Closely related to the imperative to look at one's naked flesh is an attempt to instill desire in others. Despite the protestors who insist that there are no sexual undertones to their protests, the very act of disrobing is seen as a sexual cue. This is partly because one rarely encounters nakedness outside of intimate relationships. Even within family settings nakedness retains a sense of taboo among those who are not sexually involved. Elsewhere I have discussed how revealing clothing can function as an index of sexuality; as clothing is removed, one appears to be more sexually expressive and complete nudity is the limit point at which one's sexuality is on full display.[30]

Roland Barthes may disagree with this analysis, considering his assertion that in striptease, "Woman is desexualized at the very moment when she is stripped naked."[31] Susan Stanton likewise suggests that rather than being intertwined, the act of nudity serves instead to reduce eroticism: "There is no mystery in a naked breast; there is no need to fantasize about what is beneath the clothing. It is right out there, for anyone to see. It is just a breast. The same is true for penises. . . . The experience removes your curiosity about them."[32] But despite the acquisition of knowledge concerning the other's body, the emotion in the viewer is in no way reduced. Even Barthes concedes that in the case of amateur striptease, the "'beginners' undress in front of a few hundred spectators without resorting or resorting very clumsily to magic, which unquestionably restores to the spectacle its erotic power."[33] This authenticity seems to be a fundamental basis of power for these amateurs who place themselves on display. In all of the cases examined, I have avoided the professional models, actors, and actresses because we expect that they would be comfortable in various states of public undress. We do not expect ordinary people to disrobe. But because these are everyday people, they may appear to be more accessible, more like us, more real. Moreover, because they maintain an aesthetic of the mundane, with the disrobing taking place not in a studio or the boudoir scenes of *Playboy*, but rather in public protest scenes, at shopping malls, or in the home, the reality is heightened even more.

But this incitement of desire may be counterproductive as a form of protest. Ruth Barcan argues that

> The almost total incorporation of female nudity within the erotic means that female nakedness is not all that available as a weapon of resistance, subversion, and scarcely at all as one of aggression. While a naked protestor does have a powerful repertoire of historical meanings of female nudity on which to draw, she may also have to struggle to keep unwanted meanings at bay. A young woman baring her breast or genitals at a man in an act of defiance is more likely to be leered at than feared, though perhaps she might have more luck with her buttocks. An older woman, on the other hand, is more likely to be considered mad.[34]

This was readily seen in the case of PETA and their young, attractive, chained protestors. The nudity of the individual overshadowed the message that they wished to convey. Moreover, if one wishes to be taken seriously, there must be some sense of equality in the transaction. By willingly placing oneself in a disadvantaged position, this may engender a sense of solidarity with the disadvantaged group one wishes to speak on behalf of, but it does little to help in the transaction with other, clothed individuals. Despite the prevalence of the notion that sex sells, there remains the question of *what* sex sells. In his meta-analysis of research in sex in advertising, Tom Reichert found that sex sells in cases where the product is congruent with sexual aims.[35] In essence, sex sells *sex*, which may be why nude protest sometimes is overshadowed by the erotic overtones produced in the observer. As John Ince notes, "In a culture in which figleafing is common and casual social nudity rare, most people see a naked adult body only during part of a sexual interaction, such as during lovemaking, or viewing a strip show or sexual media. In such circumstances sexual arousal accompanies the sight of genitals. Repeated over and over, that experience will help imprint an erotic program triggered just by the sight of genitals."[36]

For the protesters there is always a sense of risk in exposure, as evidenced in some of the College Humor photos that pleaded with the gatekeepers to find their images worthy of publishing. Placing oneself on display is not only an act of inciting a desire but a hope that one *can* incite desire. In other words, there is an implicit hope that he or she will be *worth looking at*. This seems particularly the case for women. Laura Sullivan argues that "the objectification of women's faces and bodies pervasive in mainstream mass media forms one of the cornerstones of women's oppression. Women internalize this objectification and in turn feel bad about ourselves (how we *look* gets translated into who we *are* and how much (self) *worth we* have), because no woman can ever measure up to the stereotypes promoted by mass mediated images."[37] Susan Bordo echoes this sentiment: "When women sense that they are not being assessed sexually—for example, as we age, or if we are disabled—it may feel like we no longer exist."[38] But this concern is salient even for women of merely average attractiveness. Evolutionary psychologists S. Craig Roberts, Emily Miner, and Todd Shackelford state that "Media exposure produces contrast effects such that, for example, a man's judgment of the attractiveness and desirability of an average woman declines."[39] In other words, the media environment provides many examples of highly attractive people, raising the bar of what is considered beautiful.

Men are also not immune from this need to measure up.[40] When naked, men expose their most (psychologically) vulnerable organ—the penis. As countless advertisements for "male enhancement" pills, penis pumps, and even surgery that severs the suspensory ligament to lengthen the penis illustrate, men experience considerable angst surrounding penis size.[41] When clothed, the length, girth, and shape of the penis are safely covered. But the penis is not the only source of male anxiety. Like women, the entire body is held up for scrutiny. Bordo notes that "the athletic, muscular male body that Calvin [Klein] plastered all over buildings, magazines, and subway stops has become an aesthetic norm, for straights as well as gays."[42] Rosalind Gill makes a similar argument: "It is not simply that there are more images of men circulating, but that a specific kind of representational practice has emerged for depicting the male body: namely an idealized and eroticized aesthetic showing a toned, young body," however, she notes that "a highly specific set of modes of representing the male body have emerged—which are quite different from sexualized representations of women's bodies."[43] All of this to say that when stripped of our protective garments, both men and women are vulnerable and risk being measured and found wanting.

Fear Me

Although this may seem counterintuitive when juxtaposed against desire and scopophilia, the naked body issues a warning that the individual who engages in such action is not to be trifled with. There is, to some extent, the notion that the individual who chooses to transgress social norms in such a dramatic way may be somewhat unstable. The taboo against public nudity is quite strong, so much so that the *Diagnostic and Statistical Manual of Mental Disorders* defines exhibition as a paraphilia.[44] These individuals have already exhibited disregard for other norms; what is left to keep them from behaving erratically, from approaching the viewer, or from otherwise acting crazy? This combination of unpredictability and disregard for commonly held norms is likely to elicit fear in others. Perhaps this is why Bataille argues, "Stripping naked is seen in civilizations where the act has full significance if not as a simulacrum of the act of killing, at least as an equivalent shorn of gravity."[45]

Because of strong societal taboos, encountering an unexpected unclothed body provides a sense of shock. I qualify this claim with "unexpected" because despite the taboo on public nudity there are still places in which one would fully expect to find nudity, such as a strip club or in a public shower at a gym or a high school. However, these spaces are limited and serve mainly to illustrate the tight control that society exerts in how one displays particular parts of his or her body. Even the common term for the genitals—privates—denotes this imperative to keep certain parts of one's body private and out of the public sphere. According to William Stephens, "It appears that notions of decent body-covering seldom extended beyond hiding the genitals, until sometime after the advent of the high civilizations,"[46] so the genitals have long been held as private even when the rest of the body could be exposed. Therefore, not all parts of the body are created equally in terms of what they communicate. While the exposure of a woman's breasts can be seen as somewhat harmless fun or playful transgression, the exposure of one's genitals is particularly shocking. The shock may come largely because the genitals are almost universally equated with sexual function, despite the fact that—or perhaps because—they also serve an excretory function. But still, the underlying message here is that one who exposes his or her body is willing to go further than others in the pursuit of their goals, whatever they may be. The viewer clearly recognizes that such relentlessness is not to be underestimated.

Yet the viewer may also fear the naked body of the other not for any physical harm that he or she may inflict, but rather the desire that may be evoked. As mentioned above, the belief that the nude body is inherently sinful remains widespread. Moreover, if the lust induced by the sight of the unclothed body takes place unconsciously and against one's will, the viewer will have lost control of the interaction. Thus, one may fear both the instability of the other, as well as the other's wantonness, which may induce undesired lust in the viewer.

Believe Me

Although some of these groups and individuals have taken the idea that "sex sells" to heart, for groups like PETA and WNBR, such an approach may be problematic—not only because of the nudity itself but also because it reveals the impulse to frame their movements as something that must be sold to the masses. William Gamson notes, "There is a well-laid cultural trap into which movement activists sometimes fall. They frame their primary task as marketing a product for consumers through the mass media . . . The logic of this approach leads one to look for a more effective marketing strategy, expressed through catchy symbols that will tap an emotional hot button and trigger the desired response."[47] In their attempt to induce belief, the protestors may simply induce attention, but such attention may not lead to the desired outcomes.

When one disrobes for a cause, they quite literally put themselves on the line. Thus, there can be little doubt of the sincerity of the rhetor's convictions. After all, with the exception of professional activists, there may be significant barriers to engaging in any protest action and few incentives.[48] Bert Klandermans suggests that participants in social movements weigh the costs and benefits of participation and that "participation is seen as a rational choice in the situation as the person perceives it, and as a way to obtain desired outcomes. In the eyes of the participant, participation is a means of reaching valued goals."[49] David Meyer counters this, arguing that one cannot simply look at protest from an economically rational perspective of benefits, noting that for many activists, "acting on behalf of one's deeply held moral and ethical beliefs is a good in itself."[50] But not all forms of participation are equal and participants in social movements may engage in a range of activities.[51] There is a significant difference between participating in a march and participating in a march *naked*. To participate in the latter requires a different level of commitment and the stronger the belief, the more likely that one will engage in more extreme forms of protest.[52] As Klandermans and Oegema note, "motivation and barriers interact to activate participation. The more people are motivated, the higher the barriers they can overcome."[53] In the face of potential embarrassment or other sanctions, the viewer can easily deduce the strength of belief one must have to publicly display his or her body for a particular cause.

The Transaction Between the Viewer and the Nude

Up to this point, I have considered how the denuded body itself communicates in a more abstract sense, but the body is not the only agent in the rhetorical transaction. Thomas Benson explains that "rhetorical being, knowing, and doing are simultaneous and overlapping actions that together constitute rhetorical action: discourse is fabricated, judgments are made, understandings are shared, agents move others and are themselves moved to belief and action, and identities are revealed and created."[54] Rhetoric is not one-directional. The person who disrobes may have some motives for doing so that go beyond what the body communicates. Moreover, the viewer's understanding of the situation will always be colored by his or her own beliefs and experiences regardless of the rhetor's intentions. Michael McGee argues that audiences for rhetorical discourse bring together the arguments and fragments of their own experience to create a complete rhetorical text.[55] Thus the rhetorical transaction is dependent upon how the viewer is likely to define the experience. This is one reason why it is so difficult to differentiate nudity from sexuality; for most individuals, the only experiences they have with nudity involve bathing or sexual behavior and neither act generally takes place in public. When such acts do take place in public, it is often for the erotic charge that comes with breaking commonly held taboos.[56] Thus, once the viewer is brought into the rhetorical equation, we are left with the possibility that the viewers will simply think that the rhetor is engaging in exhibition for his or her own sexual gratification.

The viewer is an integral part of the equation from the perspective of the nude individual as well. According to Sophie Tamas, "How we are seen haunts us. We are particularly shaped by the ways we see ourselves and internalize the gaze of others."[57] When one disrobes, he or she recognizes its power to both shock and titillate. When one seeks attention, there can be no doubt that disrobing is an effective means of garnering that attention. This was the argument made by the PETA protestors who stated that few would listen without the value-add of the naked spectacle. Others in this book likewise noted the power to gain attention through strategic exposure, but all had different expressed motives for doing so, which illustrates Tamas's point. PETA, in the case of the Running of the Nudes, described themselves as "hotties" and "babes," clearly connecting the protest to their own perceived sexual appeal.[58] That members of PETA see themselves as sexually attractive and even in some cases superior to their meat eating counterparts can be seen in campaigns such as their attempted super bowl ad "Veggie Love" and vegan babes and anti-rodeo campaign "Nobody Likes an Eight Second Ride." Thus there seems to be a need to demonstrate their sexuality in the context of their politics. The cases of the World Naked Bike Ride and the College Humor contest are likewise straightforward in their understanding that nudity can garner attention, although the WNBR has less emphasis on the appearance of the participants than PETA or College Humor. In the case of the College Humor participants sexual desirability is an essential component. Perhaps they have internalized too well the idea that their value comes from external rather than internal beauty. Theirs is clearly a world in which their "to-be-looked-at-ness" is of supreme importance.

The lactivists are a bit more complicated in how they view themselves and how they are seen, and this informs how one views them. The lactivists both exploit the power of the exposed breast while attempting to deny that power. They are also caught in a paradoxical situation of exposing a portion of their body that is viewed as erotic in a particularly non-erotic manner. None of these women express a desire to be seen as undesirable, however, and this haunts how one views them. The commonly held cultural understanding that a woman should be both desirable and, for lack of a better term, ladylike, requires that the woman keep exposure to a minimum while managing exposure to appear desirable. Thus it is understandable that many would see public breastfeeding as a transgression of both of these codes. This is not the low cut dress designed to accentuate cleavage so as to draw the eye to the breasts, inducing desire, despite the fact that more breast may be exposed in a bathing suit than in an instance of breastfeeding. Nor is it considered appropriate to expose one's breasts without an ulterior motive of drawing sexual attention to oneself. But even when the goal is to gain sexual attention, the woman who exposes too much—a skirt that is too short or a blouse that exposes too much cleavage—is held up for ridicule as a slut or as trashy. The practical considerations of feeding the child are somehow held as less important than maintaining these cultural codes.

In all of these transactions there is a sense of intimacy that has been imposed upon the viewer; the mystery of the other's flesh has now been overcome. Herein lies the power of disrobing—the viewer has little choice but to know the body of the other more completely. The illusion of civility has been cast off with the clothing and the transaction takes on a sense of immediacy that would not have been there with both participants clothed. Moreover, as mentioned above, this places the viewer in an uncomfortable position concerning his or her own morality. Joseph Libertson writes, "Nudity is a manifestation which is 'too manifest,' an appearance of that which should not have appeared—an impudency which shocks, and a vulnerability which inspires a pity that becomes desire."[59] With the intrusion of nudity, this desire becomes troubling, doubly so because of the taboos surrounding nudity and sexual desire. If the viewer finds the nude individual sexually arousing, he or she may be left with a sense of guilt over the lust that may arise as the eye lingers on the other's flesh.

But with this imposed intimacy also comes a sense of otherness. Raymond Corbey examined colonial postcards of bare-breasted African women produced between 1900 and 1930 that reflect "an erotic interest from the side of the male, European spectators," suggesting that the postcards "are visual narratives, telling a story of civilization and primitiveness, of wildness and control, of Europe and its others, of European beliefs, fears and, especially, desire."[60] It seems that in the case of public nudity, a similar impulse may be at work as the viewer looks on as the civilized one in the presence of a savage who does not understand the need for clothing. This is particularly salient for groups like PETA who symbolically become more animalistic and for members of the WNBR who seemingly reject technological progress in favor of bicycles. As such, viewing the nude can function as a way to inscribe difference and a sense of superiority over the other.

The interplay between the nude individual and the viewer may encompass endless possible reactions and reactions to reactions. But one thing seems clear: One cannot simply reduce the message to its propositional content. When the message is given by an individual at varying stages of undress, the message is affected. Although many have dismissed nudity as a means of gaining attention, it is clear that the unclothed body functions rhetorically in ways that the clothed body cannot.

The Unclothed Body as Symbolic

The undressed body is far from unproblematic in its meaning. One can easily observe that a body is uncovered to some degree, but to understand what that body means in the moment—what is signified by the nudity—is an entirely different endeavor. For example, one cannot even definitively state whether the individual is naked or nude. Kenneth Clark argues that "to be naked is to be deprived of our clothes, and the word implied some of the embarrassment most of us feel in that condition. The word 'nude,' on the other hand, carries, in educated usage, no uncomfortable overtone."[61] John Berger, on the other hand, contests this notion: "To be naked is to be oneself. To be nude is to be seen naked by others and yet not be recognized for oneself. A naked body has to be seen as an object in order to become a nude. . . . Nakedness reveals itself. Nudity is placed on display."[62] But in either of these perspectives it is difficult to define the experience of the individuals described in the previous case studies. Although they expose themselves, they do so seemingly without embarrassment, making them a nude. Yet they maintain their individuality—they are, above all, human bodies mobilized for a cause, making them naked according to Berger. But many still become objects (or at least have their identity obscured), as in the case of PETA, or in the case of the WNBR they become merely one of a mass of people, a part of a living nude sculpture. One could say that in all of these cases, the individuals are seen naked by others but not recognized for oneself.

Even deciding what to call the act performed by these individuals can seem vexing. Thus it should come as little surprise that there are a variety of interpretations for these acts, ranging from sexual exhibitionism to performances by eccentric activists to harmless fun. The range of interpretations that can be found in non-verbal communication can be almost limitless, but in this section I will discuss some of these meanings as found in the previous chapters. I suggest that nakedness is seen variously as: natural, exposure, harmlessness, difference, and erotic power.

Nakedness as Natural

As Job observed, "Naked came I out of my mother's womb, and naked shall I return thither."[63] Everyone is born naked; others clothe us, then we learn to clothe ourselves. But the unclothed body is in its true natural state. PETA used this notion to great effect by symbolically becoming more like the animals they chose to represent. WNBR participants made explicit the connection between eschewing cars and being a part of nature. The lactivists argued that breastfeeding was a natural expression of motherhood and a necessary part of feeding their children. As one woman stated, "It just seems so natural to me. This is what our breasts are for. They treated it like it was some kind of perverse act. They should be ashamed of themselves."[64]

The corollary to this sense of nakedness is that it can act as an expression of purity. In the case of the lactivists, such claims are implicit in their arguments of simply feeding their children. They complain that those around them are looking for perversion in what should be seen as a natural expression of loving motherhood. In other words, it is not their fault that others cannot see breastfeeding as natural, but rather a culture that sexualizes and commodifies the breast that is at fault. They lament that breasts are used more to sell beer and automobiles than to feed children, arguing that the socially constructed notion of the breast has overshadowed its biological function. They raise the question of why nudity itself is seen as erotic or pornographic and point an accusing finger back at the accuser. As Paul wrote to Titus, "Unto the pure all things are pure: but unto them that are defiled and unbelieving is nothing pure; but even their mind and conscience is defiled."[65]

Nakedness as Vulnerability/Exposure

The members of the World Naked Bike Ride use nudity as a metaphor for their exposure to both potential death by automobile and the exposure to environmental toxins created by automobiles. PETA likewise uses exposure as a way to illustrate the exploitation of animals. In their nudity the protestors are completely exposed to the elements and to the view of others. In short, they are without protection.

To appear naked in public is to appear at one's most vulnerable. It is unsurprising that many have had the alarming dream of being naked in public. Freud calls the interpretation of dreams the "*via regia* to a knowledge of the unconscious element in our psychic life."[66] In his treatment of such dreams, Freud attributes their interpretation to memories in the dreamer's earliest childhood, "a time when we were seen by our relatives, as well as by strange nurses, servants and visitors, in a state of insufficient clothing, and at that time we were not ashamed of our nakedness," a time without shame to which we wish to return.[67] But this sense of childhood innocence also comes with a sense of powerlessness. What one fears in public nudity is the lack of not only physical protection that comes with clothing but also the psychological protection clothing offers. While naked, it is more difficult to manage one's projected identity. The naked body is left at the mercy of others for interpretation.

Nakedness as Harmlessness

Daniel Garrison notes that even for the ancient Greeks "nakedness was normally tantamount to helplessness, representing deprivation of even the lowest common denominator of culture."[68] This is evident in PETA's use of nudity in the protest of circus animals. The WNBR made explicit use of this meaning in their protest: "Nudity is essential to show we're unarmed."[69] Another WNBR organizer echoes this connection, stating that "going naked is more peaceful than throwing bricks or starting fires."[70]

By its very nature the naked body demonstrates a lack of concealment. This makes clear that the body is all that the person has. But whether the body itself is harmless is subject to interpretation. Those who oppose actions such as the WNBR and, ironically, lactivism express concern over such displays on impressionable children. Thus the nude body cannot (usually) effectively hide a weapon. However, others may perceive the body itself as the weapon.

Nakedness as Difference

Despite the fact that all of us are naked under our clothes, very few choose to disrobe in public. Such behavior marks those individuals as significantly different from the rest of society. There is the tacit understanding that these individuals have the liberty of removing their clothing, of denouncing societal norms. All of the protestors examined here had the requisite time and means to express themselves by disrobing. Whether this is by choice or circumstance, the individuals seem to have less to lose by doing so, marking this as a strategy to be employed mainly by the indigent or those secure in their station in society. As Anita Allen notes, "an upper middle-class person can afford to care about the privacy of her body. She does not have to take a job as a stripper, whereas a poor, uneducated person might."[71] The decision to appear nude is influenced by the choices one has available. A homeless individual, for example, has little choice concerning the privacy of his or her body. The people in the cases described have employed their nudity strategically rather than out of necessity.

In their performance of difference, those who display their bodies in public make specific arguments concerning society as well as themselves. For themselves, they may demonstrate a total commitment to their cause, as argued in the case of PETA. In other words, one should be willing to completely disregard one's own body in the defense of animals. A similar case is at work in the case of the lactivists. The role of "mother" completely engulfs the previously held identity of "woman." Where a woman may not be willing to display her breasts in public, a mother recognizes that a crying child—her child—must be fed, regardless of time, place, or social squeamishness. The College Humor contest participants make arguments for society while embodying the kind of person that they see as socially desirable. Where the lactivists disavow any eroticism in their displays, the College Humor posters acknowledge their erotic power while performing a sexually liberated identity that they may or may not agree with. In all of these cases their nudity silently states, "Because of who I am, I reject the traditional norms of body display."

Nakedness as Erotic Power

This may seem obvious, but as I have noted, some of the groups here downplay the connection between exposure and sexuality, such as the WNBR and the lactivists. In fact, their cause depends completely on denying this connection. Some, however, willingly embrace the idea that nudity is inextricably entwined with the erotic. PETA tries to have it both ways—they hide their exposure behind a greater cause of helping animals. To them, the connection between nudity and sexuality is a way to draw in the masses by appealing to their prurient interests. The WNBR frames their protest in terms of fun, but some note that the secondary goal of the ride is to promote body freedom and acceptance. As one rider noted, "All bodies are beautiful."[72] For these groups, then, the nudity is simply a means to an end and any erotic charge is unavoidable in a society that sexualizes the nude body.

But to others, such as the College Humor participants, this erotic power is the desired end itself. Their participation is not merely an attempt to add their voice to a decidedly strange political poll, but rather a search for validation that their bodies are worth looking at. Some also explicitly make the connection between desirability and power, with such comments as "Does my size count for 2 votes?!" They seem invested in the idea that their bodies can sway others, take for granted that their bodies will do so, and seek to use that desire for political ends.

Can the Body Ever Lose Its Shock Value?

There have been cultures in which bare breasts and even complete nudity were seen as normal.[73] For example, Linda Boynton Arthur notes, "When missionary women arrived in Hawaii in 1820, indigenous Hawaiian women were generally bare-breasted; they wore the *pa'u* of *kapa* cloth, made from bark of the mulberry tree, and leis around the head and neck."[74] This, of course, is not the case in Western culture. With the exception of certain prescribed occasions, such as bathing or being examined by a medical doctor, exposure is often conflated with the erotic. Ince suggests that exposure reduces the mystery of the body, arguing that "in a community where family nudity is common and fig-leafing rare, genitals and female breasts are observed in contexts involving little or no sexual arousal, such as while making meals or playing games. The greater our exposure to genitals and breasts outside an erotic context, the less likely that the sight of such organs will sexually excite us. Any titillation caused by nudity is not an innate response to the sight of erotic organs, but rather a by-product of social fig-leafing."[75]

However, there seems to be more to this issue than simply the quantity and context of exposure. One element that cannot be overlooked is religious beliefs concerning the body. These beliefs vary widely, even in Christianity, although the general view seems to be that the body in inherently sinful, partly because of the potential to induce lustful thoughts and feelings in others.[76] Such strongly held values are difficult to alter or overcome, especially when they are part of the status quo. Advertisers, for example, walk the line of what is taboo, careful to avoid crossing it. They are adept at providing just enough stimulation to be naughty (e.g., women in bikinis or men with six-pack abs) without showing too much. As Plato observed, if one is to persuade, he or she must know the various kinds of souls in the audience.[77]

As with religion, one cannot ignore the media landscape in which we live. Brian McNair describes the rise of "striptease culture," a media environment in which "public nakedness, voyeurism and sexualized looking are permitted, indeed encouraged, as never before."[78] McNair describes the nude calendar produced in 1999 by the British Women's Institute (made popular by the film *Calendar Girls*), stating that "the calendar secured voluminous press coverage for its sexually transgressive qualities while signaling, perhaps more than any other single example of striptease culture in the 1990s, the nonchalance with which mainstream society—even its more conservative elements—now viewed public displays of nakedness."[79] As one news story put it,

> To be shocking now requires "supernakedness," which is how one reviewer described performance artist Annie Sprinkle's show A Public Cervix Announcement. . . . As if to acknowledge that the coyness of the traditional nudie calendar is no longer effective or eye-catching, Eizo, a German medical imaging company, released an x-ray pin-up calendar in June. Each month a woman is

> posed provocatively, but the only thing visible is skeletal structure and high heels. Truly revealing, but not very sexy.[80]

In a culture in which upskirt photos and nipple slips are *de rigueur* for maintaining celebrity status, perhaps simple nudity itself is no longer enough to shock. As Laura Verdi suggests, "Thanks to technology, the physical body has been liberated from the power of the social body: the former has been deified in order to make the latter a symbolic object of undisputed media power."[81]

Yet despite the seeming increase in frequency of public nudity, the body seems unlikely to lose its ability to shock. Rob Cover provides one potential explanation when he suggests that "a sexualization of the public sphere destabilizes the contexts in which non-sexual nakedness and gazing have been legitimated in modernity," arguing that "where there is an increased bodily and psychic anxiety about the self naked under the gaze of others, this is not merely the result of contemporary cultural codes of the obscene, vulnerability and personal privacy, but of the 'postmodern' destabilization of contexts, frames and reading practices which formerly 'protected' the naked in certain sites from slipping into significations of the sexual."[82] Even in expected situations, one may still find public nudity uncomfortable both as an observer and a participant, partly because of the coding of the nude body as intrinsically sexual.

But it seems that the coding of nudity has, in certain contexts, added a dimension besides erotic display—that of nudity as playful fun. Groups such as the WNBR explicitly connect their nude protest with fun. Others have also drawn on the idea that disrobing is simply a matter of having fun. In the summer of 2006, a young woman decided to sit naked on a park bench in Brattleboro, Vermont. The *Washington Post* reported, "In July, half a dozen young people bared their bodies in a parking lot encircled by the backs of bookstores, coffee shops and restaurants. 'It's just an act of freedom,' said Adhi Palar, 19. 'We're just doing so because we can.'"[83] It seems that the amusement factor comes not only from the nudity itself, but also from the shock value of the spectacle. Such discourses of freedom and fun are common in slight transgressions such as skinny dipping and are not a recent phenomenon. Such amusement from the shock of nudity has been part of popular culture as far back as Allen Funt's 1970 film *What Do You Say to a Naked Lady*, which documented ordinary people's reactions to unexpected nudity. But as Bataille notes, "Often the transgression of a taboo is no less subject to rules than the taboo itself."[84] The youth in Brattleboro seem to be pushing the limits of the traditionally prescribed locations of public nudity, such as nude beaches and skinny dipping, but there are limits concerning what they can actually *do* in such a state of nudity. They are not completely free to use their bodies as they wish; masturbating or engaging in sexual activity in public would be met with swift arrest and punishment. But actions such as playing guitar in the park or sitting on a bench can be viewed (by some) with detached bemusement by the rest of society as the foibles of youth.

"The body" cannot be easily constructed as a coherent idea in which all bodies are the same. Obviously, Western society views male and female toplessness differently, for example.[85] It is not enough to simply lump all bodies together as "humans" or "people." Rather, it seems that somewhere along the line we have made a fundamental shift in how we view the body, from viewing the body as a body to viewing the body as a symbol. In other words, the body has become an object of desire rather than simply a person. All of this is wrapped up in a particular performance of self. One cannot simply *be* anymore, if we ever could, because of the symbolicity of the body. As Elizabeth Grosz observes, "The body is a peculiar 'thing,' for it is never quite reducible to being merely a thing; nor does it ever quite manage to rise above the status of thing."[86] Thus we reach the question of what we are trying to be when we disrobe, but we cannot avoid the question of what it means to inhabit a body. Because the first question has been addressed above, I will explore the latter question here.

As physical bodies, we continually communicate, whether by movement, voice, or adornment. Watzlawick, Bavelas and Jackson suggest that "one cannot *not* communicate," concluding that "others, in turn, cannot *not* respond to these communications and are thus themselves communicating."[87] Such communication can be directed at others as well as the self. For example, Sharon Phelan argues, "High heels make a woman feel prettier. When a woman feels prettier, it alters her self-image in a social, emotional, and psychosensuous sense. She is actually a different person in mood and self-projection."[88] To be embodied is to experience the world through the senses. To wear high heels is to feel the reshaping of the legs, to feel the difference in one's gait, to see the world from a slightly higher perspective. When one stands closer to another person, they will feel different, whether that feeling comes from the discomfort of standing so near a stranger or the excitement of standing near someone he or she finds desirable. Either way, he or she will experience a physiological response to the other's presence. In short, when one uses the body, he or she will generally alter individual bodily experience.

In addition to experiencing the world, one also becomes a part of the world that others perceive. This is crucial when we consider how people perform identity. One may adorn oneself with particular clothing to signal religious or cultural affiliation, to display a desired identity, or to signal sexual availability. One may attempt to portray a particular message though posture, gesture, facial expression, or tone. But this notion of being in the world as a part of another's world is an important facet of embodiment. One always feels the gaze of the other, whether it is an actual other, an imagined other that he or she will encounter throughout the day, or simply the desire to conform to the vision one has of him or herself. One cannot escape the performance of embodiment because we are all trapped within our own flesh and no matter how hard we try to not communicate anything, our bodies will always betray us.

To perform nude embodiment is to make explicit the performance of self. When clothed, one can hide behind social norms that have varying latitudes of acceptance for transgression. As a man, I may wear a skirt; although my choice of clothing may appear suspect, no one will question my choice of wearing clothing. When one strips off his or her clothing, they experience embodiment in its most concrete form. At that moment, there is nothing but the body present. Still, removing clothing changes the experience of being in one's body just as putting on high heels or wearing a corset will. If one disrobes outside, he or she is quite exposed to the elements. If it is cold, he or she may begin to involuntarily shiver and feel the tightening of the skin and the raising of goose bumps. To become naked is to simultaneously lose control of one's body while asserting the agency to do so. Thus, disrobing is a forceful statement of will. But the body communicates beyond the physiological attributes mentioned. Wearing clothing draws one's attention to specific places on the body. For example, it is difficult to avoid reading a message on someone's shirt. A brightly colored bikini draws the eye to the breasts, pubic area, and buttocks. Ince notes that "covering specific zones of the body sets them apart; the naked body cannot be so neatly compartmentalized."[89] Thus, a nude body is an individual quantity rather than an amalgam of body parts set apart by various articles of clothing.

We are left, then, with the body as body and the body as symbol, but the two are interrelated. If symbols are created by people, they can change the meanings of those symbols through language. This, more than anything, is the work of rhetoric. Feminist scholars have noted that language affects women's self image; Victoria Pitts writes, "Reclaiming or resistance ideology implies that social inscriptions on the body can be rewritten, and the body—especially the female genitals and breasts—can be reclaimed."[90] What all of these cases suggest is a suspicion that by disrobing one can reconfigure cultural symbols, whether the disrobing is integral to the desired change (lactivism) or not (WNBR, College Humor, PETA).

But what would such a reconstruction look like? In other words, with what would we replace the currently existing symbolic construction of the nude body? This is a difficult question to answer because different groups employ nudity to different ends with different assumptions concerning the body. The lactivists are invested in maintaining the exposed body as neutral, as merely a natural state of being. PETA and WNBR rely on the shocking nature of the exposed body as a means to garner attention to their cause, but also use the nude body as a metaphor for solidarity with the animals or exposure to potential danger. The body resists neatly circumscribed constellations of meanings even within these limited case studies. Like all forms of symbolism, including language itself, the body is polysemic. Herein lies its power to induce both discomfort and arousal. One thing is certain, however: So long as the naked body remains taboo, it will be viewed with an uncomfortable mixture of disgust at those uncouth enough to expose it and desire for those that do.

Notes

1. Debra Hawhee, "Rhetorics, Bodies, and Everyday Life," *RSQ: Rhetoric Society Quarterly* 36, no. 2 (2006): 157.

2. Parke G. Burgess, "The Rhetoric of Moral Conflict: Two Critical Dimensions," *Quarterly Journal of Speech* 56, no. 2 (1970): 120.

3. Edwin Black, "The Second Persona," *Quarterly Journal of Speech* 56 (1970): 119.

4. William H. Masters and Virginia E. Johnson, *Human Sexual Inadequacy* (Boston: Little, Brown and Company, 1970), 24.

5. Susie Bright, *Full Exposure: Opening up to Sexual Creativity and Erotic Expression* (San Francisco: HarperSanFrancisco, 1999), 120.

6. Georg Feuerstein, *Sacred Sexuality: Living the Vision of the Erotic Spirit* (Los Angeles, CA: Jeremy P. Tarcher, 1992), 16.

7. Gen. 2:25.

8. Gen. 3:7-11.

9. Gen. 3:21.

10. Gen. 3:22.

11. "In the Study," *Expository Times* 34, no. 9 (1923): 420.

12. Melvyn Matthews, "'Who Told You That You Were Naked?' A Metaphor for a Christian Theology of Atonement," *Theology* 93, no. 753 (1990): 220.

13. James David Velleman, "The Genesis of Shame," *Philosophy & Public Affairs* 30, no. 1 (2001): 31.

14. Peter Matheson, "The Hammer, the Sickle and the Rainbow: Sixteenth-Century Radicalism and Its Interpreters," *Theology* 93, no. 751 (1990): 21.

15. William G. Gray, *Temple Magic: Building the Personal Temple: Gateway to Inner Worlds* (St. Paul, MN: Llewellyn Publications, 1988), 113.

16. However, there are many readings of this Biblical narrative. For one particularly interesting reading which places the story of the Garden of Eden within its cultural context and attempts to re-envision the role of Eve, see Judith E. McKinlay, "Bothering to Enter the Garden of Eden Once Again," *Feminist Theology* 19, no. 2 (2011): 143-53.

17. Masters and Johnson, *Human Sexual Inadequacy*, 254. They also make a connection between religious orthodoxy and vaginismus, a condition in which involuntary contractions make vaginal sexual intercourse painful or impossible. See pp. 254-56.

18. Augustine, *The City of God*, trans. Gerald G. Walsh, Demetrius B. Zema, Grace Monahan, and Daniel Honan (Garden City, NY: Image Books, 1958), 315.

19. Jürgen Habermas, *Moral Consciousness and Communicative Action*, trans. Christian Lenhardt and Shierry Weber Nicholsen (Cambridge, MA: MIT Press, 1990), 108.

20. Aristotle, "Rhetoric," in *The Complete Works of Aristotle: The Revised Oxford Translation*, ed. Jonathan Barnes, 2152-269 (Princeton, NJ: Princeton University Press, 1984), 1356a14-15.

21. Mitchell Berbrier, "From Logos to Pathos in Social Psychology and Academic Argumentation: Reconciling Postmodernism and Positivism in a Sociology of Persuasion," *Argumentation* 11, no. 1 (1997): 47.

22. See Roland Barthes, *Image, Music, Text*, trans. Stephen Heath (New York: Hill and Wang, 1977), 148; Michael Calvin McGee, "Text, Context, and the Fragmentation of Contemporary Culture," *Western Journal of Communication* 54, no. 3 (1990): 274-89.

23. For example, one component of the limbic system in particular, the amygdala, acts as a sort of "emergency switch." Daniel Goleman, *Emotional Intelligence* (New York: Bantam Books, 1995), 22-26. For more on this line of argument, see Brett Lunceford, "The Science of Orality: Implications for Rhetorical Theory," *Review of Communication* 7, no. 1 (2007): 83-102.

24. Aristotle, "Rhetoric," 1356a4-11.

25. Gary Alan Fine, "Public Narration and Group Culture: Discerning Discourse in Social Movements," in *Social Movements and Culture*, ed. Hank Johnston and Bert Klandermans, 127-43 (Minneapolis: University of Minnesota Press, 1995), 131.

26. See W. L. Marshall, Krista Payne, H. E. Barbaree, and A. Eccles, "Exhibitionists: Sexual Preferences for Exposing," *Behaviour Research and Therapy* 29, no. 1 (1991): 37-40.

27. Amanda Nosko, Eileen Wood, and Serge Desmarais, "Unsolicited Online Sexual Material: What Affects Our Attitudes and Likelihood to Search for More?" *Canadian Journal of Human Sexuality* 16, no. 1-2 (2007): 7. However, there is more to this than simply gender differences. Another study suggests that how one views sexual appeals is not only a factor of gender, but also one's attitudes toward sexuality. Women who held more conservative attitudes were likely to view gratuitous sexual appeals negatively and women who held liberal attitudes toward sex were likely to view such ads positively. For stimulus material, they used an advertisement featuring nudity. See Jaideep Sengupta and Darren W. Dahl, "Gender-Related Reactions to Gratuitous Sex Appeals in Advertising," *Journal of Consumer Psychology* 18, no. 1 (2008): 73.

28. See Nikki Craft, "PETA: Where Only Women Are Treated Like Meat," www.nostatusquo.com/ACLU/PETA/peta.html; Maneesha Deckha, "Disturbing Images: PETA and the Feminist Ethics of Animal Advocacy," *Ethics & the Environment* 13, no. 2 (2008): 35-76; Laura K. Hahn, "I'm Too Sexy for Your Movement: An Analysis of the Failure of the Animal Rights Movement to Promote Vegetarianism," in *Arguments About Animal Ethics*, ed. Greg Goodale and Jason Edward Black, 79-95 (Lanham, MD: Lexington Books, 2010); kitchenMage, "Dear PETA, Women Are Animals Too . . . ," blog .kitchenmage.com/2007/10/dear-peta-women.html.

29. Ann R. Fischer, Sonya K. Bettendorf, and Yu-Wei Wang, "Contextualizing Sexual Objectification," *The Counseling Psychologist* 39, no. 1 (2011): 128.

30. Brett Lunceford, "Smeared Makeup and Stiletto Heels: Clothing, Sexuality, and the Walk of Shame," in *College Sex: Philosophy for Everyone: Philosophers with Benefits*, ed. Robert Stewart and Michael Bruce, 51-60 (Hoboken, NJ: Wiley-Blackwell, 2010). For more on the idea of indexicality, see Charles S. Peirce, *Collected Papers*, ed. Charles Hartshorne, and Paul Weiss, 6 vols. (Cambridge, MA: Belknap Press of Harvard University Press, 1960), 2:248.

31. Roland Barthes, *Mythologies*, trans. Annette Lavers (New York: Hill and Wang, 1972), 84.

32. Susan L. Stanton, *Being Naked: Attitudes Toward Nudity Through the Ages* (St. Clair Shores, MI: Ablaze Press, 2001): 9.

33. Barthes, *Mythologies*, 86.

34. Ruth Barcan, *Nudity: A Cultural Anatomy* (New York: Berg, 2004), 197-98.

35. Tom Reichert, "Sex in Advertising Research: A Review of Content, Effects, and Functions of Sexual Information in Consumer Advertising," *Annual Review of Sex Research* 13 (2002): 241-73.

36. John Ince, *The Politics of Lust* (Amherst, NY: Prometheus Books, 2005), 30.

37. Laura L. Sullivan, "Cyberbabes: (Self-) Representation of Women and the Virtual Male Gaze," *Computers and Composition* 14, no. 2 (1997): 192.

38. Susan Bordo, *The Male Body: A New Look at Men in Public and in Private* (New York, NY: Farrar, Straus and Giroux, 1999), 173.

39. S. Craig Roberts, Emily J. Miner, and Todd K. Shackelford, "The Future of an Applied Evolutionary Psychology for Human Partnerships," *Review of General Psychology* 14, no. 4 (2010): 319.

40. For more on male body dissatisfaction, see Marika Tiggemann, Yolanda Martins, and Libby Churchett, "Beyond Muscles: Unexplored Parts of Men's Body Image," *Journal of Health Psychology* 13, no. 8 (2008): 1163-72.

41. For more on penis enhancement products and surgeries, see L. Janáčková, P. Weiss, and M. Krejča, "Psychological Aspects of Penoplasty," *Sexologies* 18, no. 3 (2009): 186-88; Chi-Ying Li, Oliver Kayes, Phillip D. Kell, Nim Christopher, Suks Minhas, David J. Ralph, "Penile Suspensory Ligament Division for Penile Augmentation: Indications and Results," *European Urology* 49, no. 4 (2006): 729-33; Yoram Vardi, Yaron Harshai, Tamir Gil, and Ilan Gruenwald, "A Critical Analysis of Penile Enhancement Procedures for Patients with Normal Penile Size: Surgical Techniques, Success, and Complications," *European Urology* 54, no. 5 (2008): 1042-50; Emily Wentzell and Jorge Salmerón, "You'll 'Get Viagraed': Mexican Men's Preference for Alternative Erectile Dysfunction Treatment," *Social Science & Medicine* 68, no. 10 (2009): 1759-65.

42. Bordo, *The Male Body*, 185.

43. Rosalind Gill, "Beyond the 'Sexualization of Culture' Thesis: An Intersectional Analysis of 'Sixpacks,' 'Midriffs' and 'Hot Lesbians' in Advertising," *Sexualities* 12, no. 2 (2009): 143, 147-48.

44. American Psychiatric Association, *Diagnostic and Statistical Manual of Mental Disorders: DSM-IV-TR*, 4th Text Rev. ed. (Washington, DC: American Psychiatric Association, 2000), section 302.4. One may note that the diagnostic criteria include: "Over a period of at least 6 months, recurrent, intense sexually arousing fantasies, sexual urges, or behaviors involving the exposure of one's genitals to an unsuspecting stranger." It is the "or" that places many of these protest actions into the realm of the seemingly pathological.

45. Georges Bataille, *Erotism: Death & Sensuality*, trans. Mary Dalwood (San Francisco: City Lights Books, 1986), 18.

46. William N. Stephens, "A Cross-Cultural Study of Modesty," *Cross-Cultural Research* 7, no. 1 (1972): 5.

47. William A. Gamson, "Constructing Social Protest," in *Social Movements and Culture*, ed. Hank Johnston and Bert Klandermans, 85-106 (Minneapolis: University of Minnesota Press, 1995), 105.

48. For more on the professionalization of social movements, see Kevin Djo Everett, "Professionalization and Protest: Changes in the Social Movement Sector, 1961-1983," *Social Forces* 70, no. 4 (1992): 957-75; John D. McCarthy and Mayer N. Zald, "The Trend of Social Movements in America: Professionalization and Resource Mobilization," in *Social Movements in an Organizational Society: Collected Essays*, ed. Mayer N. Zald and John D. McCarthy, 337-91 (New Brunswick, NJ: Transaction Books, 1987); Suzanne Staggenborg, "The Consequences of Professionalization and Formalization in the Pro-Choice Movement," *American Sociological Review* 53, no. 4 (1988): 585-605.

49. Bert Klandermans, "Mobilization and Participation: Social-Psychological Expansisons of Resource Mobilization Theory," *American Sociological Review* 49, no. 5 (1984): 597.

50. David S. Meyer, *The Politics of Protest: Social Movements in America* (New York: Oxford University Press, 2007), 50.

51. Steven E. Barkan, Steven F. Cohn, and William H. Whitaker, "Beyond Recruitment: Predictors of Differential Participation in a National Antihunger Organization," *Sociological Forum* 10, no. 1 (1995): 113-34.

52. See M. Kent Jennings and Ellen Ann Andersen, "Support for Confrontational Tactics among AIDS Activists: A Study of Intra-Movement Divisions," *American Journal of Political Science* 40, no. 2 (1996): 311-34.

53. Bert Klandermans and Dirk Oegema, "Potentials, Networks, Motivations, and Barriers: Steps Towards Participation in Social Movements," *American Sociological Review* 52, no. 4 (1987): 520.

54. Thomas W. Benson, "Rhetoric as a Way of Being," in *American Rhetoric: Context and Criticism*, ed. Thomas W. Benson, 293-322 (Carbondale: Southern Illinois University Press, 1989), 318.

55. McGee, "Text, Context," 274-89.

56. See David Bell, "Bodies, Technologies, Spaces: On 'Dogging,'" *Sexualities* 9, no. 4 (2006): 387-407; Corie J. Hammers, "Making Space for an Agentic Sexuality? The Examination of a Lesbian/Queer Bathhouse," *Sexualities* 11, no. 5 (2008): 547-72; David Redmon, "Playful Deviance as an Urban Leisure Activity: Secret Selves, Self-Validation, and Entertaining Performances," *Deviant Behavior* 24, no. 1 (2003): 27-51.

57. Sophie Tamas, "Sketchy Rendering: Seeing an Other," *Qualitative Inquiry* 15, no. 3 (2009): 615.

58. People for the Ethical Treatment of Animals, "Running of the Nudes: Out with the Old-in with the Nude! // Next Year's Race // Look Who's Running," www.runningofthenudes.com/look_whos_running.asp.

59. Joseph Libertson, *Proximity, Levinas, Blanchot, Bataille, and Communication* (The Hague: Martinus Nijhoff, 1982), 223.

60. Raymond Corbey, "Alterity: The Colonial Nude," *Critique of Anthropology* 8, no. 3 (1988): 75.

61. Kenneth Clark, *The Nude: A Study in Ideal Form* (Princeton, NJ: Princeton University Press, 1972), 3.

62. John Berger, *Ways of Seeing* (London: British Broadcasting Corporation; Penguin, 1972), 54.

63. Job 1:21. See also, Eccl. 5:15.

64. Quoted in Margie Boule, "No Need for Tantrum over Nursing of Baby," *Oregonian* (Portland, OR), February 12, 1998, C1.

65. Titus 1:15.

66. See Sigmund Freud, "The Interpretation of Dreams," in *The Basic Writings of Sigmund Freud*, ed. A. A. Brill, 177-549 (New York: Modern library, 1938), 540.

67. Freud, "The Interpretation of Dreams," 294.

68. Daniel H. Garrison, *Sexual Culture in Ancient Greece* (Norman, OK: University of Oklahoma Press, 2000), 179.

69. Julie Pace, "Protestors Shun Bike Shorts to Bare Views on Oil, Cars," *Tampa Tribune* March 13, 2006, 6.

70. Quoted in Shelby Martin, "Cyclists in Bike Parade to Air Views in Ride of Protest," *Register-Guard* (Eugene, OR), March 14, 2009, A1.

71. Anita L. Allen, "Coercing Privacy," *William and Mary Law Review* 40, no. 3 (1999): 736.

72. Gale Hayley, "Bare Biking a Real Sea Change," *Nelson Mail* (Nelson, NZ), March 15, 2010, 3.

73. William Stephens provides a listing of groups that range from complete nakedness to covering only the genitals to covering the whole body with a range of covering in between. Stephens notes, however, that many of the groups he cites "have now been Christianized and wear more clothes than they did in precontact time." See Stephens, "A Cross-Cultural Study of Modesty," 3-5.

74. Linda Boynton Arthur, "Cultural Authentication Refined: The Case of the Hawaiian Holokū," *Clothing and Textiles Research Journal* 15, no. 3 (1997): 129.

75. Ince, *The Politics of Lust*, 30-31.

76. Much of this enmity toward the body stems from the Judeo-Christian religious tradition, but others have different conceptions of the nude body. For example, in his discussion of magic rituals, William Gray notes that "Complete nudity for ceremonial practice is the oldest possible style dating from our most primitive past and is still in use among Pagans, though far from being very widely popular." Jo Pearson likewise notes that Wiccans employ ritual nudity, as well as sexual practices. Gray, *Temple Magic*, 113; Jo Pearson, "Inappropriate Sexuality? Sex Magic, S/M and Wicca (or 'Whipping Harry Potter's Arse!')" *Theology & Sexuality* 11, no. 2 (2005): 31-42.

77. Plato, "Phaedrus," in *The Collected Dialogues of Plato, Including the Letters*, ed. Edith Hamilton and Huntington Cairns, 475-525 (Princeton, NJ: Princeton University Press, 1961), 271d.

78. Brian McNair, *Striptease Culture: Sex, Media and the Democratization of Desire* (London: Routledge, 2002), ix.

79. McNair, *Striptease Culture*, 96. For more on the use of nudity in charity calendars, see Pamela Turton-Turner, "The Role of Ridicule in Naked Charity Calendars," *Continuum: Journal of Media & Cultural Studies* 21, no. 3 (2007): 419-32.

80. Ryan Bigge, "Being Naked in Public Has a Number of Spiritual, Cultural and Political Implications," *Toronto Star* August 8, 2010, IN2.

81. Laura Verdi, "The Symbolic Body and the Rhetoric of Power," *Social Analysis* 54, no. 2 (2010), 111.

82. Rob Cover, "The Naked Subject: Nudity, Context and Sexualization in Contemporary Culture," *Body & Society* 9, no. 3 (2003): 55.

83. Lisa Rathke, "Teens Test Naked City's Limits; Town Asks: Should There Be a Law?" *Washington Post* September 24, 2006, A8.

84. Bataille, *Erotism*, 65.

85. See Virginia F. Milstead, "Forbidding Female Toplessness: Why Real Difference Jurisprudence Lacks Support and What Can Be Done About It," *University of Toledo Law Review* 36, no. 2 (2005): 273-320; Danielle Moriber, "A Right to Bare All? Female Public Toplessness and Dealing with the Laws That Prohibit," *Cardozo Public Law, Policy, and Ethics Journal* 8, no. 2 (2010): 453-84. These scholars argue that differences in legal treatment of female toplessness rests almost entirely on a heterosexual male view of the breast. Yet this difference based on sex is problematic. Luke Boso argues that the courts' definition of sex as "a biological predetermination" and that "'real differences' between males and females justify the prohibition of female toplessness" leads to absurd results in the case of transsexuals. Luke Boso, "A (Trans)Gender-Inclusive Equal Protection Analysis of Public Female Toplessness," *Law & Sexuality: A Review of Lesbian, Gay, Bisexual and Transgender Legal Issues* 18, no. 1 (2009): 146. The question of how to classify transsexuals and what legal rights should be afforded them has seemed particularly vexing not only for the United States, but for other countries as well. See Lori Chambers, "Unprincipled Exclusions: Feminist Theory, Transgender Jurisprudence, and Kimberly Nixon," *Canadian Journal of Women & the Law* 19, no. 2 (2007): 305-34;

Robyn Emerton, "Finding a Voice, Fighting for Rights: The Emergence of the Transgender Movement in Hong Kong," *Inter-Asia Cultural Studies* 7, no. 2 (2006): 243-69; Sheila Jeffreys, "They Know It When They See It: The UK Gender Recognition Act 2004," *British Journal of Politics & International Relations* 10, no. 2 (2008): 328-45; Laura H. Norton, "Neutering the Transgendered: Human Rights and Japan's Law No. 111," *Georgetown Journal of Gender & the Law* 7, no. 2 (2006): 187-216; Stephen Whittle, "The Opposite of Sex Is Politics—The UK Gender Recognition Act and Why It Is Not Perfect, Just Like You and Me," *Journal of Gender Studies* 15, no. 3 (2006): 267-71. Moreover, this is troubled even further in the distinction between sex and gender which, although often used interchangeably, are not synonymous concepts. Some scholars argue that sex is a biological indicator while gender is socially constructed, for example, Brooks Aylor and Marianne Dainton, "Biological Sex and Psychological Gender as Predictors of Routine and Strategic Relational Maintenance," *Sex Roles* 50, no. 9/10 (2004): 689-97; Mary Riege Laner, "A Rejoinder to Mark Creswell," *Sociological Inquiry* 73, no. 1 (2003): 152-56; Mary Riege Laner, "'Sex' Versus 'Gender': A Renewed Plea," *Sociological Inquiry* 70, no. 4 (2000): 462-74. However, others argue that sex itself is not simply biologically, but also socially constructed. See Sandra Lipsitz Bem, "Dismantling Gender Polarization and Compulsory Heterosexuality: Should We Turn the Volume Down or Up?" *Journal of Sex Research* 32, no. 4 (1995): 329-34; Judith Butler, *Bodies That Matter: On the Discursive Limits of 'Sex'* (New York: Routledge, 1993); Iris Marion Young, *On Female Body Experience: "Throwing Like a Girl" and Other Essays* (New York: Oxford University Press, 2005). The categorization of sex becomes even more complicated in the case of intersex individuals that do not neatly fall into the traditional binary of male or female. See Ellen K. Feder and Katrina Karkazis, "What's in a Name? The Controversy over 'Disorders of Sex Development,'" *Hastings Center Report* 38, no. 5 (2008): 33-36; Brendan Gough, Nicky Weyman, Julie Alderson, Gary Butler, and Mandy Stoner, "'They Did Not Have a Word': The Parental Quest to Locate a 'True Sex' for Their Intersex Children," *Psychology & Health* 23, no. 4 (2008): 493-507; Claudia Lang and Ursula Kuhnle, "Intersexuality and Alternative Gender Categories in Non-Western Cultures," *Hormone Research* 69, no. 4 (2008): 240-50; Iain Morland, "'The Glans Opens Like a Book': Writing and Reading the Intersexed Body," *Continuum: Journal of Media & Cultural Studies* 19, no. 3 (2005): 335-48. As such, it is difficult to definitively make judgments based on one's gender or sex because these may be called into question, regardless of how obvious they may seem.

86. Elizabeth A. Grosz, *Volatile Bodies: Toward a Corporeal Feminism* (Crows Nest, NSW: Allen & Unwin, 1994), xi.

87. Paul Watzlawick, Janet Beavin Bavelas, and Don D. Jackson, *Pragmatics of Human Communication: A Study of Interactional Patterns, Pathologies, and Paradoxes* (New York: W. W. Norton & Company, 1967), 49.

88. Sharon T. Phelan, "Fads and Fashions: The Price Women Pay," *Primary Care Update for OB/GYNS* 9, no. 4 (2002): 140.

89. Ince, *Politics of Lust*, 19.

90. Victoria L. Pitts, "'Reclaiming' the Female Body: Embodied Identity Work, Resistance and the Grotesque," *Body & Society* 4, no. 3 (1998): 71. See also Petula Sik Ying Ho and Adolf Ka Tat Tsang, "Beyond the Vagina-Clitoris Debate—From Naming the Genitals to Reclaiming the Woman's Body," *Women's Studies International Forum* 28, no. 6 (2005): 523-34; Brett Lunceford, "The Walk of Shame: A Normative Description," *ETC: A Review of General Semantics* 65, no. 4 (2008): 319-29.

Bibliography

"600 Strip Naked on Glacier in Global Warming Protest: Chilling Message from Wear-Nothing Activists to Do-Nothing Politicians." Greenpeace, www.greenpeace.org/international/en/news/features/naked-glacier-tunick-08182007/.

"A Simple Policy." *Ottawa Citizen* January 14, 2009, A14.

"About FFF." Fuck for Forest, www.fuckforforest.com/en/about.html.

Adler, Amy. "Symptomatic Cases: Hysteria in the Supreme Court's Nude Dancing Decisions." *American Imago* 64, no. 3 (2007): 297-316.

Adler, Libby S. "A Short Essay on the Baring of Breasts." *Harvard Women's Law Journal* 23 (2000): 219-26.

Aguirre, B. E., E. L. Quarantelli, and Jorge L. Mendoza. "The Collective Behavior of Fads: The Characteristics, Effects, and Career of Streaking." *American Sociological Review* 53, no. 4 (1988): 569-84.

"All the Nudes That's Fit to Print." *Coventry Evening Telegraph* June 23, 2006.

Allain, Annelies, and Kean Yeong Joo. "The Youngest Market: Baby Food Peddlers Undermine Breastfeeding." *Multinational Monitor* 29, no. 1 (2008): 17-22.

Allen, Anita L. "Disrobed: The Constitution of Modesty." *Villanova Law Review* 51, no. 4 (2006): 841-58.

American Psychiatric Association. *Diagnostic and Statistical Manual of Mental Disorders: DSM-IV-TR*. 4th Text Rev. ed. Washington, DC: American Psychiatric Association, 2000.

Anderson, Alex K., Grace Damio, David A. Himmelgreen, Yu-Kuei Peng, Sofia Segura-Pérez, and Rafael Pérez-Escamilla. "Social Capital, Acculturation, and Breastfeeding Initiation among Puerto Rican Women in the United States." *Journal of Human Lactation* 20, no. 1 (2004): 39-45.

Anderson, Judith L. "Breast, Hips, and Buttocks Revisited: Honest Fatness for Honest Fitness." *Ethology and Sociobiology* 9, no. 5 (1988): 319-24.

Anderson, William A. "The Social Organization and Social Control of a Fad: Streaking on a College Campus." *Journal of Contemporary Ethnography* 6, no. 2 (1977): 221-40.

Andrews, Donnovan. "Bullish Steps." *Black Enterprise* May 2002, 114.

"Angry Breast-Feeding Mother Organizes Starbucks 'Nurse-in.'" *Columbian* (Vancouver, WA) August 10, 2004, A8.

AnimaNaturalis. "Protesta "Sin Piel" En Madrid Diciembre De 2009." www.animanaturalis.org/v/548.

Appleton, Dorothy. "Lady Godiva." *Dawn*, no. 36 (2000): 12.

Ardener, Shirley. "A Note on Gender Iconography: The Vagina." In *The Cultural Construction of Sexuality*, edited by Patricia Caplan, 113-42. London: Routledge, 1987.

Arditi, Lynn. "Nursing Mom's Protest at Mall No Secret." *Providence Journal* July 2, 2006, B1.

Aristotle. *On Rhetoric: A Theory of Civic Discourse*. Translated by George Alexander Kennedy. New York: Oxford University Press, 1991.

———. "Rhetoric." In *The Complete Works of Aristotle: The Revised Oxford Translation*, edited by Jonathan Barnes, 2152-269. Princeton, NJ: Princeton University Press, 1984.

Arpan, Laura M., Kaysee Baker, Youngwon Lee, Taejin Jung, Lori Lorusso, and Jason Smith. "News Coverage of Social Protests and the Effects of Photographs and Prior Attitudes." *Mass Communication & Society* 9, no. 1 (2006): 1-20.

Arthur, Linda Boynton. "Cultural Authentication Refined: The Case of the Hawaiian Holokū." *Clothing and Textiles Research Journal* 15, no. 3 (1997): 129-39.

Asen, Robert. "A Discourse Theory of Citizenship." *Quarterly Journal of Speech* 90, no. 2 (2004): 189-211.

Atkins-Sayre, Wendy. "Articulating Identity: People for the Ethical Treatment of Animals and the Animal/Human Divide." *Western Journal of Communication* 74, no. 3 (2010): 309-28.

———. "Protection from 'Animal Rights Lunatics': The Center for Consumer Freedom and Animal Rights Rhetoric." In *Arguments About Animal Ethics*, edited by Greg Goodale and Jason Edward Black, 147-62. Lanham, MD: Lexington Books, 2010.

"Assembly Panel Again Rejects Public Breast-Feeding Proposal." *San Francisco Chronicle* May 11, 1995, A20.

Associated Press. "Airline Acts on Protest by Nursing Mother; Punishes Attendant Who Ordered Her Off." *The Record* (Bergen County, NJ) November 18, 2006, A8.

Auer, Holly. "Mothers Stage Nurse-in to Support Breast-Feeding." *Post and Courier* (Charleston, SC) July 1, 2005, B1.

Augustine. *The City of God*. Translated by Gerald G. Walsh, Demetrius B. Zema, Grace Monahan and Daniel Honan. Garden City, NY: Image Books, 1958.

Ayalah, Daphna, and Isaac J. Weinstock, eds. *Breasts: Women Speak About Their Breasts and Their Lives*. New York: Summit Books, 1979.

Aylor, Brooks, and Marianne Dainton. "Biological Sex and Psychological Gender as Predictors of Routine and Strategic Relational Maintenance." *Sex Roles* 50, no. 9/10 (2004): 689-97.

Baer, Edward C. "Promoting Breastfeeding: A National Responsibility." *Studies in Family Planning* 12, no. 4 (1981): 198-206.

Bailey, Cathryn. "We Are What We Eat: Feminist Vegetarianism and the Reproduction of Racial Identity." *Hypatia* 22, no. 2 (2007): 39-59.

Bakardjieva, Maria. *Internet Society: The Internet in Everyday Life*. London: Sage, 2005.

Baker, Courtney. "Emmett Till, Justice, and the Task of Recognition." *Journal of American Culture* 29, no. 2 (2006): 111-24.

Bakhtin, Mikhail M. *Rabelais and His World*. Translated by Hélène Iswolsky. Bloomington: Indiana University Press, 1984.

Barcan, Ruth. "Female Exposure and the Protesting Woman." *Cultural Studies Review* 8, no. 2 (2002): 62-82.

———. "Home on the Rage: Nudity, Celebrity, and Ordinariness in the Home Girls/Blokes Pages." *Continuum: Journal of Media & Cultural Studies* 14, no. 2 (2000): 145-58.

———. *Nudity: A Cultural Anatomy*. New York: Berg, 2004.

Bare Witness. "Bare Witness Peace." www.barewitness.org.

Barkan, Steven E., Steven F. Cohn, and William H. Whitaker. "Beyond Recruitment: Predictors of Differential Participation in a National Antihunger Organization." *Sociological Forum* 10, no. 1 (1995): 113-34.

Barker, James R. "Tightening the Iron Cage: Concertive Control in Self-Managing Teams." *Administrative Science Quarterly* 38, no. 3 (1993): 408-37.

Barker-Plummer, Bernadette. "News as a Political Resource: Media Strategies and Political Identity in the U.S. Women's Movement, 1966-1975." *Critical Studies in Mass Communication* 12, no. 3 (1995): 306-24.

Barron, Chris. "A Big Question for Gay Republicans." *Gay & Lesbian Review Worldwide* 11 (2004): 27-29.

Barry, Dan. "In Times Sq. These Days, Naked Ladies Have Causes." *New York Times* November 23, 2005, B2.

Barry, Rick. "Oh, Sure: Now We're Having a Nude Bike-In." *Tampa Tribune* January 23, 2005, 1.

Barthes, Roland. *Camera Lucida: Reflections on Photography*. Translated by Richard Howard. New York: Hill and Wang, 1981.

———. *Image, Music, Text*. Translated by Stephen Heath. New York: Hill and Wang, 1977.

———. *Mythologies*. Translated by Annette Lavers. New York: Hill and Wang, 1972.

Baruh, Lemi. "Publicized Intimacies on Reality Television: An Analysis of Voyeuristic Content and Its Contribution to the Appeal of Reality Programming." *Journal of Broadcasting & Electronic Media* 53, no. 2 (2009): 190-210.

Bataille, Georges. *Erotism: Death & Sensuality*. Translated by Mary Dalwood. San Francisco: City Lights Books, 1986.

———. *The Tears of Eros*. San Francisco: City Lights Books, 1989.

Baudrillard, Jean. *Simulacra and Simulation*. Ann Arbor: University of Michigan Press, 1994.

Beetles, Andrea C., and Lloyd C. Harris. "Consumer Attitudes Towards Female Nudity in Advertising: An Empirical Study." *Marketing Theory* 5 (2005): 397-432.

Bell, David. "Bodies, Technologies, Spaces: On 'Dogging.'" *Sexualities* 9, no. 4 (2006): 387-407.

Bell, David, and Ruth Holliday. "Naked as Nature Intended." *Body & Society* 6, no. 3 (2000): 127-40.

Bem, Sandra Lipsitz. "Dismantling Gender Polarization and Compulsory Heterosexuality: Should We Turn the Volume Down or Up?" *Journal of Sex Research* 32, no. 4 (1995): 329-34.

Benford, Robert D, and David A Snow. "Framing Processes and Social Movements: An Overview and Assessment." *Annual Review of Sociology* 26 (2000): 611-39.

Benhabib, Seyla. "Models of Public Space: Hannah Arendt, the Liberal Tradition, and Jürgen Habermas." In *Habermas and the Public Sphere*, edited by Craig J. Calhoun, 73-98. Cambridge, MA: MIT Press, 1992.

Benson, Thomas W. "Rhetoric as a Way of Being." In *American Rhetoric: Context and Criticism*, edited by Thomas W. Benson, 293-322. Carbondale: Southern Illinois University Press, 1989.

Berbrier, Mitchell. "From Logos to Pathos in Social Psychology and Academic Argumentation: Reconciling Postmodernism and Positivism in a Sociology of Persuasion." *Argumentation* 11, no. 1 (1997): 35-50.

Berger, John. *Ways of Seeing*. London: British Broadcasting Corporation; Penguin, 1972.

Berger, Lawrence R. "When Should One Discourage Breast-Feeding." *Pediatrics* 67, no. 2 (1981): 300-02.

Berger, Peter L., and Thomas Luckmann. *The Social Construction of Reality: A Treatise in the Sociology of Knowledge*. New York: Anchor Books, 1966.

Bernhardt, Colette. "The Guide: Events: The Guide: World Naked Bike Ride Brighton, Cardiff, London, Sheffield." *The Guardian* June 13, 2009, 42.

Bernstein, Mary. "The Analytic Dimensions of Identity: A Political Identity Framework." In *Identity Work in Social Movements*, edited by Jo Reger, Daniel J. Myers and Rachel L. Einwohner, 277-301. Minneapolis: University of Minnesota Press, 2008.

"Bicyclists Ride in Protest, and in Little Else." *New York Times* June 12, 2005, 1.9.

Bigge, Ryan. "Being Naked in Public Has a Number of Spiritual, Cultural and Political Implications." *Toronto Star* August 8, 2010, IN.2.

Bitzer, Lloyd F. "Aristotle's Enthymeme Revisited." *Quarterly Journal of Speech* 45, no. 4 (1959): 399-408.

Black, Edwin. "The Second Persona." *Quarterly Journal of Speech* 56 (1970): 109-19.

Black, Jason Edward. "Extending the Rights of Personhood, Voice, and Life to Sensate Others: A Homology of Right to Life and Animal Rights Rhetoric." *Communication Quarterly* 51, no. 3 (2003): 312-31.

Bonar, S. "Putting Breastfeeding on the Agenda." *RCM Midwives* 10, no. 1 (2007): 13.

"Boosting Natural Feeding." *Midwifery Matters*, no. 111 (2006): 24-25.

Bordo, Susan. *The Male Body: A New Look at Men in Public and in Private*. New York: Farrar, Straus and Giroux, 1999.

Boris, Eileen, and S. J. Kleinberg. "Mothers and Other Workers: (Re)Conceiving Labor, Maternalism, and the State." *Journal of Women's History* 15, no. 3 (2003): 90-117.

Bosmajian, Haig A. "Defining the 'American Indian': A Case Study in the Language of Suppression." *Speech Teacher* 22, no. 2 (1973): 89-99.

———. "Obscenity and Protest." *Today's Speech* 18 (1970): 9-14.

Boso, Luke. "A (Trans)Gender-Inclusive Equal Protection Analysis of Public Female Toplessness " *Law & Sexuality: A Review of Lesbian, Gay, Bisexual and Transgender Legal Issues* 18, no. 1 (2009): 143-62.

Boule, Margie. "No Need for Tantrum over Nursing of Baby." *Oregonian* (Portland, OR) February 12, 1998, C1.

Braman, Sandra, and Stephanie Roberts. "Advantage ISP: Terms of Service as Media Law." *New Media & Society* 5, no. 3 (2003): 422-48.

"Breastfeeding-Related Maternity Practices at Hospitals and Birth Centers—United States, 2007." *MMWR. Morbidity And Mortality Weekly Report* 57, no. 23 (2008): 621-25.

"Breastfeeding Should Not Be Shunned." *The Sun* June 25, 2007, 1.

Bright, Susie. *Full Exposure: Opening up to Sexual Creativity and Erotic Expression*. San Francisco: HarperSanFrancisco, 1999.

Bronstein, Carolyn. "Representing the Third Wave: Mainstream Print Media Framing of a New Feminist Movement." *Journalism & Mass Communication Quarterly* 82, no. 4 (2005): 783-803.

Brown, Gavin, and Jenny Pickerill. "Space for Emotion in the Spaces of Activism." *Emotion, Space and Society* 2, no. 1 (2009): 24-35.

Bryant, Salatheia. "Moms Meet at Galleria for `Nurse-in' Protest." *Houston Chronicle* July 2, 2004, 27.

Buodo, Giulia, Michela Sarlo, and Daniela Palomba. "Attentional Resources Measured by Reaction Times Highlight Differences within Pleasant and Unpleasant, High Arousing Stimuli." *Motivation & Emotion* 26, no. 2 (2002): 123-38.

Burgess, Parke G. "The Rhetoric of Moral Conflict: Two Critical Dimensions." *Quarterly Journal of Speech* 56, no. 2 (1970): 120-30.

Burke, Kenneth. *A Grammar of Motives*. New York: Prentice-Hall, 1945.

———. *A Rhetoric of Motives*. New York: Prentice-Hall, 1950.

Butler, Judith. *Bodies That Matter: On the Discursive Limits of "Sex."* New York: Routledge, 1993.

———. *Gender Trouble: Feminism and the Subversion of Identity*. New York: Routledge, 1990.

Caballero, Marjorie J., James R. Lumpkin, and Charles S. Madden. "Using Physical Attractiveness as an Advertising Tool: An Empirical Test of the Attraction Phenomenon." *Journal of Advertising Research* 29, no. 4 (1989): 16-22.

Cafazzo, Debbie. "Lactation Intolerance." *News Tribune* (Tacoma, WA) September 24, 2006, E1.

Calhoun, Craig. "Social Theory and the Politics of Identity." In *Social Theory and the Politics of Identity*, edited by Craig J. Calhoun, 9-36. Oxford: Blackwell, 1994.

Carpenter, Mackenzie. "Bottle vs. Breastfeeding New Products Aim to Make Nursing Less Stressful, but Cultural Confusion Engulfs Moms No Matter Which Method Is Used." *Pittsburgh Post-Gazette* October 31, 2006, C1.

Carr-Gomm, Philip. *A Brief History of Nakedness*. London: Reaktion, 2010.

Chaiken, Shelly. "Communicator Physical Attractiveness and Persuasion." *Journal of Personality & Social Psychology* 37, no. 8 (1979): 1387-97.

Chambers, Lori. "Unprincipled Exclusions: Feminist Theory, Transgender Jurisprudence, and Kimberly Nixon." *Canadian Journal of Women & the Law* 19, no. 2 (2007): 305-34.

Charland, Maurice. "Constitutive Rhetoric: The Case of the *Peuple Quebecois*." *Quarterly Journal of Speech* 73, no. 2 (1987): 133-50.

Chu, Richard. "Residents Walk, Run, Cycle Nude to Raise $150,000, Awareness." *Vancouver Sun* June 13, 2005, B1.

Church, Jill Howard. "A Whopper of a Champaign." *The Animals' Agenda* 21, no. 3 (2001): 10.

Chvasta, Marcyrose. "Anger, Irony, and Protest: Confronting the Issue of Efficacy, Again." *Text & Performance Quarterly* 26, no. 1 (2006): 5-16.

Cialdini, Robert B. *Influence: The Psychology of Persuasion*. Rev. ed. New York: Morrow, 1993.

Cialdini, Robert B., and Noah J. Goldstein. "Social Influence: Compliance and Conformity." *Annual Review of Psychology* 55, no. 1 (2004): 591-621.

Cicero, Marcus Tullius. *De Oratore, Books I-II*. Translated by E. W. Sutton and H. Rackham. Cambridge, MA: Harvard University Press, 1942.

Cixous, Hélène. "The Laugh of the Medusa." In *The Rhetorical Tradition: Readings from Classical Times to the Present*, edited by Patricia Bizzell and Bruce Herzberg, 1524-36. Boston, MA: Bedford/St. Martin's, 2001.

Clark, John, and Nuno Themudo. "The Age of Protest: Internet-Based 'Dot-Causes' and the 'Anti-Globalization' Movement." In *Globalizing Civic Engagement: Civil Soci-*

ety and Transnational Action, edited by John Clark, 109-26. London: Earthscan Publications, 2003.

Clark, Kenneth. *The Nude: A Study in Ideal Form*. Princeton, NJ: Princeton University Press, 1972.

Clifford, J., and E. McIntyre. "Who Supports Breastfeeding?" *Breastfeeding Review* 16, no. 2 (2008): 9-19.

Collins, Lois M. "Provo Woman Protests Facebook's Removal of Breast-Feeding Photo." *Deseret News* (Salt Lake City, UT) December 31, 2008, B2.

Conneller, Chantal. "Becoming Deer. Corporeal Transformations at Star Carr." *Archaeological Dialogues* 11, no. 01 (2004): 37-56.

Corbey, Raymond. "Alterity: The Colonial Nude." *Critique of Anthropology* 8, no. 3 (1988): 75-92.

Coren, Victoria. "7 Days: I Don't Normally Sign Petitions, but in This Case." *Observer* (London) January 4, 2009, 34.

Correa, Barbara. "Moms to Stage 'Nurse-in' at Applebee's Locations." *Inland Valley Daily Bulletin* (Ontario, CA) September 7, 2007.

Cottle, Simon. "Reporting Demonstrations: The Changing Media Politics of Dissent." *Media, Culture & Society* 30, no. 6 (2008): 853-72.

Cover, Rob. "The Naked Subject: Nudity, Context and Sexualization in Contemporary Culture." *Body & Society* 9, no. 3 (2003): 53-72.

Craft, Nikki. "PETA: Where Only Women Are Treated Like Meat." www.nostatusquo.com/ACLU/PETA/peta.html.

Curran, John. "'Nurse-Ins' Support Breast-Feeding Mother Who Was Ordered Off Plane." *Daily Breeze* (Torrance, CA) November 22, 2006, A11.

Currier, Joel. "Mostly Naked Cyclists Draw Crowd for Ride." *St. Louis Post-Dispatch* August 3, 2008, C2.

"Cyclists to Give Their Cause Extra Exposure." *Daily Mail* (London) June 12, 2004, 4.

Davidson, H. R. Ellis. "The Legend of Lady Godiva." *Folklore* 80, no. 2 (1969): 107-21.

Davis, Victoria. "Safety for Cyclists." *Nelson Mail* (Nelson, NZ) March 1, 2007, 9.

Davison, W. Phillips. "The Third-Person Effect in Communication." *Public Opinion Quarterly* 47, no. 1 (1983): 1-15.

De Wit, John B. F., Enny Das, and Raymond Vet. "What Works Best: Objective Statistics or a Personal Testimonial? An Assessment of the Persuasive Effects of Different Types of Message Evidence on Risk Perception." *Health Psychology* 27, no. 1 (2008): 110-15.

Debord, Guy. *The Society of the Spectacle*. Translated by Donald Nicholson-Smith. New York: Zone Books, 1994.

Deckha, Maneesha. "Disturbing Images: PETA and the Feminist Ethics of Animal Advocacy." *Ethics & the Environment* 13, no. 2 (2008): 35-76.

DeFour, Matthew. "Law to Protect Nursing Women ; County Proposal Bars Hindering Breast-Feeding." *Wisconsin State Journal* (Madison, WI) January 4, 2007, B1.

Deleuze, Gilles. "Ethology: Spinoza and Us." In *The Body: A Reader*, edited by Mariam Fraser and Monica Greco, 58-61. New York: Routledge, 2005.

Deleuze, Gilles, and Félix Guattari. *A Thousand Plateaus: Capitalism and Schizophrenia*. Translated by Brian Massumi. Minneapolis, MN: University of Minnesota Press, 1987.

Delevie, Brian, and Isshaela Ingham. "Unconscionable or Communicable: The Transference of Holocaust Photography in Cyber Space." *Afterimage* 31, no. 6 (2004): 7.

DeLuca, Kevin Michael. *Image Politics: The New Rhetoric of Environmental Activism*. Mahwah, NJ: Lawrence Erlbaum Associates, 1999.

———. "Unruly Arguments: The Body Rhetoric of Earth First!, ACT UP, and Queer Nation." *Argumentation & Advocacy* 36, no. 1 (1999): 9-21.

DeLuca, Kevin Michael, and Jennifer Peeples. "From Public Sphere to Public Screen: Democracy, Activism, and the 'Violence' of Seattle." *Critical Studies in Media Communication* 19, no. 2 (2002): 125-51.

Dewan, Shaila. "How Photos Became Icon of Civil Rights Movement." *New York Times* August 28, 2005, 1.12.

Dewey, John. *The Public and Its Problems*. Athens, OH: Swallow Press, 1991.

Dicker, Glenda. "Let the People See What I've Seen: In Praise of Mamie Till." *Southern Quarterly* 45, no. 4 (2008): 152-54.

Dietrich, Dawn. "Refashioning the Techno-Erotic Woman: Gender and Textuality in the Cybercultural Matrix." In *Virtual Culture: Identity and Communication in Cybersociety*, edited by Steve Jones, 169-84. London: Sage Publications, 1997.

Dorea, Jose G. "Maternal Smoking and Infant Feeding: Breastfeeding Is Better and Safer." *Maternal & Child Health Journal* 11, no. 3 (2007): 287-91.

Downs, Anthony. *An Economic Theory of Democracy*. New York: Harper, 1957.

Dudley, Sid C. "Consumer Attitudes toward Nudity in Advertising." *Journal of Marketing Theory and Practice* 7, no. 4 (1999): 89-95.

Dyens, Ollivier. *Metal and Flesh: The Evolution of Man: Technology Takes Over*. Cambridge, MA: MIT Press, 2001.

Eagan, Margery. "Let's Nip Nursing Protest in Bud." *Boston Herald* April 26, 2007, 6.

Eck, Beth A. "Men Are Much Harder: Gendered Viewing of Nude Images." *Gender & Society* 17, no. 5 (2003): 691-710.

Eigenhauser, George J., Jr. "Criminal Sanctions against Nude Sunbathing in California." *Criminal Justice Journal* 1, no. 3 (1978): 491-508.

Elliott, Richard, Abigail Jones, Andrew Benfield, and Matt Barlow. "Overt Sexuality in Advertising: A Discourse Analysis of Gender Responses." *Journal of Consumer Policy* 18, no. 2,3 (1995): 187-217.

Ellul, Jacques. *Propaganda: The Formation of Men's Attitudes*. Translated by Konrad Kellen and Jean Lerner. New York: Knopf, 1965.

———. "Technology and Democracy." In *Democracy in a Technological Society*, edited by Langdon Winner, 35-50. Dordrecht: Kluwer, 1992.

Elvin, John. "Tough Times for Buff Protests." *Insight on the News* 19, no. 5 (2003): 17.

Emerton, Robyn. "Finding a Voice, Fighting for Rights: The Emergence of the Transgender Movement in Hong Kong." *Inter-Asia Cultural Studies* 7, no. 2 (2006): 243-69.

"Emma to Bare All for Naked Truth over Cruelty to Bulls." *Journal* (Newcastle-upon-Tyne, UK), June 19, 2006, 7.

Esterik, Penny Van, and Ted Greiner. "Breastfeeding and Women's Work: Constraints and Opportunities." *Studies in Family Planning* 12, no. 4 (1981): 184-97.

Evans, Christopher. "Victoria's Secret Welcomes Breast-Feeding Protesters." *Plain Dealer* (Cleveland, OH) July 2, 2006, B.1.

Everett, Kevin Djo. "Professionalization and Protest: Changes in the Social Movement Sector, 1961-1983." *Social Forces* 70, no. 4 (1992): 957-75.

Facebook. "Statement of Rights and Responsibilities." www.facebook.com/#!/terms.php.

———. "Statistics." Facebook, www.facebook.com/press/info.php?statistics.

Farvid, Panteá, and Virginia Braun. "'Most of Us Guys Are Raring to Go Anytime, Anyplace, Anywhere': Male and Female Sexuality in *Cleo* and *Cosmo*." *Sex Roles* 55, no. 5 (2006): 295-310.

Feder, Ellen K., and Katrina Karkazis. "What's in a Name? The Controversy over 'Disorders of Sex Development.'" *Hastings Center Report* 38, no. 5 (2008): 33-36.

FEMEN. "Femen - The Official Web Site." www.femen.org/.

Fernback, Jan. "The Individual within the Collective: Virtual Ideology and the Realization of Collective Principles." In *Virtual Culture: Identity and Communication in Cybersociety*, edited by Steve Jones, 36-54. London: Sage Publications, 1997.

Ferree, Myra Marx. "The Rise and Fall of 'Mommy Politics': Feminism and Unification in (East) Germany." *Feminist Studies* 19, no. 1 (1993): 89-115.

Feuerstein, Georg. *Sacred Sexuality: Living the Vision of the Erotic Spirit*. 1st ed. Los Angeles, CA: Jeremy P. Tarcher, 1992.

Fine, Gary Alan. "Public Narration and Group Culture: Discerning Discourse in Social Movements." In *Social Movements and Culture*, edited by Hank Johnston and Bert Klandermans, 127-43. Minneapolis: University of Minnesota Press, 1995.

Fischer, Ann R., Sonya K. Bettendorf, and Yu-Wei Wang. "Contextualizing Sexual Objectification." *The Counseling Psychologist* 39, no. 1 (2011): 127-39.

Fisher, Walter R. *Human Communication as Narration: Toward a Philosophy of Reason, Value, and Action*. Columbia, SC: University of South Carolina Press, 1987.

———. "Narration as a Human Communication Paradigm: The Case of Public Moral Argument." *Communication Monographs* 51, no. 1 (1984): 1-22.

Fitchburg, Laura E. Olsen Dugan. "City, State Should Join County in Support of Breast-Feeding." *Madison Capital Times* (Madison, WI) January 8, 2007, A7.

Fortunati, Leopoldina. "The Human Body: Natural and Artificial Technology." In *Machines That Become Us: The Social Context of Personal Communication Technology*, edited by James Everett Katz, 71-87. New Brunswick, NJ: Transaction Publishers, 2003.

Foss, Sonja K. "Framing the Study of Visual Rhetoric: Toward a Transformation of Rhetorical Theory." In *Defining Visual Rhetorics*, edited by Charles A. Hill and Marguerite H. Helmers, 303-13. Mahwah, NJ: Lawrence Erlbaum, 2004.

———. "Theory of Visual Rhetoric." In *Handbook of Visual Communication Research: Theory, Methods, and Media*, edited by Ken Smith, Sandra Moriarty, Keith Kenney and Gretchen Barbatsis, 141-52. Mahwah, NJ: Lawrence Erlbaum, 2005.

Foucault, Michel. *Discipline and Punish: The Birth of the Prison*. Translated by Alan Sheridan. 2nd Vintage Books ed. New York: Vintage Books, 1995.

Fowler, Thomas L. "Of Moons, Thongs, Holdings and Dicta: State V. Fly and the Rule of Law." *Campbell Law Review* 22, no. 2 (2000): 253-316.

Frank III, Arthur W. "Reality Construction in Interaction." *Annual Review of Sociology* 5 (1979): 167-91.

Frank, Deborah A., Stephen J. Wirtz, James R. Sorenson, and Timothy Heeren. "Commercial Discharge Packs and Breast-Feeding Counseling: Effects on Infant-Feeding Practices in a Randomized Trial." *Pediatrics* 80, no. 6 (1987): 845.

Frank, John. "Panel Approves Breast-Feeding Bill." *Post and Courier* (Charleston, SC) February 8, 2006, B1.

Frankel, Todd C. "Activists, Circus Take Wraps Off Feud over Animals' Treatment." *St. Louis Post-Dispatch* November 17, 2005, A1.

Frantzich, Stephen E. *Cyberage Politics 101: Mobility, Technology and Democracy*. New York: Peter Lang, 2002.

Fraser, Nancy. "Rethinking the Public Sphere: A Contribution to the Critique of Actually Existing Democracy." In *Habermas and the Public Sphere*, edited by Craig J. Calhoun, 109-42. Cambridge, MA: MIT Press, 1992.

———. *Unruly Practices: Power, Discourse, and Gender in Contemporary Social Theory*. Minneapolis: University of Minnesota Press, 1989.

Frerichs, Leah, Julie L. Andsager, Shelly Campo, Mary Aquilino, and Carolyn Stewart Dyer. "Framing Breastfeeding and Formula-Feeding Messages in Popular U.S. Magazines." *Women & Health* 44, no. 1 (2006): 95-118.

Freud, Sigmund. "The Interpretation of Dreams." In *The Basic Writings of Sigmund Freud*, edited by A. A. Brill, 177-549. New York: Modern library, 1938.

Furnham, Adrian, Viren Swami, and Krupa Shah. "Body Weight, Waist-to-Hip Ratio and Breast Size Correlates of Ratings of Attractiveness and Health." *Personality and Individual Differences* 41, no. 3 (2006): 443-54.

Galtry, Judith. "Extending the 'Bright Line': Feminism, Breastfeeding, and the Workplace in the United States." *Gender and Society* 14, no. 2 (2000): 295-317.

Gamson, William A. "Constructing Social Protest." In *Social Movements and Culture*, edited by Hank Johnston and Bert Klandermans, 85-106. Minneapolis: University of Minnesota Press, 1995.

Garlick, Hattie. "The Naked and the Dead Brave." *Times* (London) June 8, 2007, 2.

Garrison, Daniel H. *Sexual Culture in Ancient Greece*. Norman: University of Oklahoma Press, 2000.

Geva, Dorit. "Not Just Maternalism: Marriage and Fatherhood in American Welfare Policy." *Social Politics: International Studies in Gender, State and Society* 18, no. 1 (2011): 24-51.

Giglia, R. C., and C. W. Binns. "Alcohol, Pregnancy and Breastfeeding; a Comparison of the 1995 and 2001 National Health Survey Data." *Breastfeeding Review* 16, no. 1 (2008): 17-24.

Giles, Fiona. "The Well-Tempered Breast: Fostering Fluidity in Breastly Meaning and Function." *Women's Studies* 34, no. 3/4 (2005): 301-26.

Gill, Rosalind. "Beyond the 'Sexualization of Culture' Thesis: An Intersectional Analysis of 'Sixpacks,' 'Midriffs' and 'Hot Lesbians' in Advertising." *Sexualities* 12, no. 2 (2009): 137-60.

Gilmartin, Shannon K. "Changes in College Women's Attitudes toward Sexual Intimacy." *Journal of Research on Adolescence* 16, no. 3 (2006): 429-54.

Gitlin, Todd. *The Whole World Is Watching: Mass Media in the Making & Unmaking of the New Left*. Berkeley: University of California Press, 1980.

Gleason, Matt. "'Naked' Ride Will Find Pedalers in Skivvies." *Tulsa World* July 23, 2009, D3.

Goffman, Erving. *Encounters: Two Studies in the Sociology of Interaction*. Indianapolis: Bobbs-Merrill, 1961.

———. *The Presentation of Self in Everyday Life*. Garden City, NY: Doubleday, 1959.

Goleman, Daniel. *Emotional Intelligence*. New York: Bantam Books, 1995.

Goodnight, G. Thomas. "Public Discourse." *Critical Studies in Mass Communication* 4, no. 4 (1987): 428-31.

Goodwin, Daisy. "A Bit Too in Your Facebook." *Sunday Times* (London) January 4, 2009, 4.

Gordon, Andrea. "Facebook Sours on Nursing Mom; More Than 10,000 So-Called 'Lactivists' Have Signed an Online Petition Protesting the Website's Policy." *Toronto Star* September 12, 2007, A1.

GoTopless.org, "National Go-Topless Day!" www.gotopless.org/.

Gough, Brendan, Nicky Weyman, Julie Alderson, Gary Butler, and Mandy Stoner. "'They Did Not Have a Word': The Parental Quest to Locate a 'True Sex' for Their Intersex Children." *Psychology & Health* 23, no. 4 (2008): 493-507.

Graham, Elaine L. *Representations of the Post/Human: Monsters, Aliens, and Others in Popular Culture*. New Brunswick, NJ: Rutgers University Press, 2002.

Grammer, Karl, Bernhard Fink, Anders P. Møller, and Randy Thornhill. "Darwinian Aesthetics: Sexual Selection and the Biology of Beauty." *Biological Reviews* 78, no. 3 (2003): 385-407.

Granovetter, Mark. "The Strength of Weak Ties: A Network Theory Revisited." *Sociological Theory* 1 (1983): 201-33.

Granovetter, Mark S. "The Strength of Weak Ties." *American Journal of Sociology* 78, no. 6 (1973): 1360-80.

Gray, William G. *Temple Magic: Building the Personal Temple: Gateway to Inner Worlds*. St. Paul, MN: Llewellyn Publications, 1988.

Gregg, Richard B. "The Ego-Function of the Rhetoric of Protest." *Philosophy & Rhetoric* 4, no. 2 (1971): 71-91.

Griskevicius, Vladas, Noah J. Goldstein, Chad R. Mortensen, Jill M. Sundie, Robert B. Cialdini, and Douglas T. Kenrick. "Fear and Loving in Las Vegas: Evolution, Emotion, and Persuasion." *Journal of Marketing Research* 46, no. 3 (2009): 384-95.

Grosz, Elizabeth A. *Volatile Bodies: Toward a Corporeal Feminism*. Crows Nest, NSW: Allen & Unwin, 1994.

Gueguen, Nicolas. "Women's Bust Size and Men's Courtship Solicitation." *Body Image* 4, no. 4 (2007): 386-90.

Gunkel, David J. *Hacking Cyberspace*. Boulder, CO: Westview Press, 2001.

Habermas, Jürgen. *Moral Consciousness and Communicative Action*. Translated by Christian Lenhardt and Shierry Weber Nicholsen. Cambridge, MA: MIT Press, 1990.

———. *The Structural Transformation of the Public Sphere: An Inquiry into a Category of Bourgeois Society*. Translated by Thomas Burger. Cambridge, MA: MIT Press, 1989.

Hague, Barry N., and Brian Loader. "Digital Democracy: An Introduction." In *Digital Democracy: Discourse and Decision Making in the Information Age*, edited by Barry N. Hague and Brian Loader, 3-22. London: Routledge, 1999.

Hahn, Laura K. "I'm Too Sexy for Your Movement: An Analysis of the Failure of the Animal Rights Movement to Promote Vegetarianism." In *Arguments About Animal Ethics*, edited by Greg Goodale and Jason Edward Black, 79-95. Lanham, MD: Lexington Books, 2010.

Haiman, Franklyn S. "The Rhetoric of the Streets: Some Legal and Ethical Considerations." *Quarterly Journal of Speech* 53, no. 2 (1967): 99-114.

Hales, Deborah Jean. "Promoting Breastfeeding: Strategies for Changing Hospital Policy." *Studies in Family Planning* 12, no. 4 (1981): 167-72.

Hammers, Corie J. "Making Space for an Agentic Sexuality? The Examination of a Lesbian/Queer Bathhouse." *Sexualities* 11, no. 5 (2008): 547-72.

Hanjani, Bahar. "The Illustration of Beauty: Super-Exposed in the U.S., Veiled in Iran." *Semiotica* 187, no. 1-4 (2011): 309-21.

Harbolt, Thomas P. "Welcome to Oregon! No "Pasties" Or "G-Strings" Required: *Barnes v. Glen Theater, Inc.*" *Willamette Law Review* 29, no. n1 (1993): 129-50.

Harmon, Amy. "'Lactivists' Taking Their Cause, and Their Babies, to the Streets." *New York Times* June 7, 2005, B3.

Harold, Christine, and Kevin Michael DeLuca. "Behold the Corpse: Violent Images and the Case of Emmett Till." *Rhetoric & Public Affairs* 8, no. 2 (2005): 263-86.

Harrison, Bridget. "Breast Unrest—Library Sorry for Tossing 'Milk Mom.'" *New York Post* October 2, 2002, 7.

Harrower, Tim. *Inside Reporting: A Practical Guide to the Craft of Journalism*. Boston: McGraw-Hill, 2007.

Hart, Anne. "Hey Facebook, Breast-Feeding Isn't Obscene." *Savannah Morning News* January 4, 2009, E1.

Hartland, E. Sidney. "Peeping Tom and Lady Godiva." *Folklore* 1, no. 2 (1890): 207-26.

Hartley, Matt. "Breastfeeding Mothers Protest Photo Ban." *Globe and Mail* (Toronto) December 31, 2008, L3.

Harvey, Mike. "Mothers Angered as Facebook Cuts 'Indecent' Images; Decision by Social Networking Website to Ban Breastfeeding Photos Sparks Outrage." *Ottawa Citizen* December 31, 2008, A7.

———. "Mothers Reveal Naked Anger as Facebook Bans 'Indecent' Images of Breast-Feeding." *Times* (London) December 31, 2008, 15.

Hauser, Gerard A. "Features of the Public Sphere." *Critical Studies in Mass Communication* 4, no. 4 (1987): 437-41.

———. "Incongruous Bodies: Arguments for Personal Sufficiency and Public Insufficiency." *Argumentation & Advocacy* 36, no. 1 (1999): 1-8.

Hausman, Bernice L. "Things (Not) to Do with Breasts in Public: Maternal Embodiment and the Biocultural Politics of Infant Feeding." *New Literary History* 38, no. 3 (2007): 479-504.

Hawhee, Debra. "Rhetorics, Bodies, and Everyday Life." *RSQ: Rhetoric Society Quarterly* 36, no. 2 (2006): 155-64.

Hay, Amy M. "Recipe for Disaster: Motherhood and Citizenship at Love Canal." *Journal of Women's History* 21, no. 1 (2009): 111-34.

Hayley, Gale. "Bare Biking a Real Sea Change." *Nelson Mail* (Nelson, NZ) March 15, 2010, 3.

Helderman, Rosalind S. "Md. Moms Say No to Coverup at Starbucks; Women Push Chain for Policy Allowing Public Breast-Feeding in All U.S. Stores." *Washington Post* August 9, 2004, B3.

Hendrie, Colin A., Helena D. Mannion, and Georgina K. Godfrey. "Evidence to Suggest That Nightclubs Function as Human Sexual Display Grounds." *Behaviour* 146, no. 10 (2009): 1331-48.

Hetsroni, Amir. "Sexual Content on Mainstream TV Advertising: A Cross-Cultural Comparison." *Sex Roles* 57, no. 3/4 (2007): 201-10.

"Hey Facebook, Breastfeeding Is Not Obscene! (Official Petition to Facebook): Info." Facebook, www.facebook.com/#!/group.php?gid=2517126532&v=info.

Hibberd, James. "Poll: TV Violence Worse Than Nudity." *Television Week* 24, no. 4 (2005): 59.

Hill, Russell A., Sophie Donovan, and Nicola F. Koyama. "Female Sexual Advertisement Reflects Resource Availability in Twentieth-Century UK Society." *Human Nature* 16, no. 3 (2005): 266-77.

Hirschfeld, Magnus. *Curious Sex Customs in the Far East: The World Journey of a Sexologist*. Translated by O. P. Green. New York: Capricorn Books, 1935.

Ho, Petula Sik Ying, and Adolf Ka Tat Tsang. "Beyond the Vagina-Clitoris Debate—From Naming the Genitals to Reclaiming the Woman's Body." *Women's Studies International Forum* 28, no. 6 (2005): 523-34.

Hoffman, Abbie. *The Best of Abbie Hoffman*. Edited by Daniel Simon. New York: Four Walls Eight Windows, 1989.

Hoffman, Merle. "Do Feminists Need to Liberate Animals, Too? Carol Adams Sees Feminism as a Visionary Philosophy That Includes Stewardship of the Earth." *On The Issues* 4, no. 2 (1995): 18-24, 54-56.

Hogan, J. Michael. *The Nuclear Freeze Campaign: Rhetoric and Foreign Policy in the Telepolitical Age*. East Lansing: Michigan State University Press, 1994.

Hollander, D. "Long-Term Breastfeeding, Especially of First Child, Lowers Breast Cancer Risk." *Family Planning Perspectives* 28, no. 5 (1996): 239.

Hopkinson, Judy M. "Response to 'Is Breast Really Best? Risk and Total Motherhood in the National Breastfeeding Awareness Campaign'." *Journal of Health Politics, Policy And Law* 32, no. 4 (2007): 637-48.

Houck, Davis W. "Killing Emmett." *Rhetoric & Public Affairs* 8, no. 2 (2005): 225-62.

Howard, Alan J. "When Can the Moral Majority Rule? The Real Dilemma at the Core of the Nude Dancing Cases." *Saint Louis University Law Journal* 44, no. 3 (2000): 897-908.

Howard, Cynthia R., Fred M. Howard, Michael Weitzman, and Ruth Lawrence. "Antenatal Formula Advertising: Another Potential Threat to Breast-Feeding." *Pediatrics* 94, no. 1 (1994): 102-4.

Huang, Grant. "Breastfeeding Moms Stage 'Nurse-in' at BWI." *Maryland Gazette* (Glen Burnie, MD) November 23, 2006, A2.

Hudis, Mark. "PETA Picks a Pic of Naked Pop Star." *MediaWeek* 5, no. 13 (1995): 44.

"In the Study." *Expository Times* 34, no. 9 (1923): 416-26.

Ince, John. *The Politics of Lust*. Amherst, NY: Prometheus Books, 2005.

Jamieson, Robert L., Jr. "Pundit Who Slipped on Mother's Milk." *Seattle Post-Intelligencer* June 28, 2005, B1.

Janáčková, L., P. Weiss, and M. Krejča. "Psychological Aspects of Penoplasty." *Sexologies* 18, no. 3 (2009): 186-88.

Jeffreys, Sheila. "They Know It When They See It: The UK Gender Recognition Act 2004." *British Journal of Politics & International Relations* 10, no. 2 (2008): 328-45.

Jennings, M. Kent, and Ellen Ann Andersen. "Support for Confrontational Tactics among AIDS Activists: A Study of Intra-Movement Divisions." *American Journal of Political Science* 40, no. 2 (1996): 311-34.

Jewkes, Yvonne, and Keith Sharp. "Crime, Deviance and the Disembodied Self: Transcending the Dangers of Corporeality." In *Dot.Cons*, edited by Yvonne Jewkes, 1-14. Portland, OR: Willan Publishing, 2003.

Jiao, C., B. Turman, P. Weerakoon, and P. Knight. "Alterations in Grip Strength During Male Sexual Arousal." *International Journal of Impotence Research* 18, no. 2 (2006): 206-09.

Johnson, Stefanie K., Kenneth E. Podratz, Robert L. Dipboye, and Ellie Gibbons. "Physical Attractiveness Biases in Ratings of Employment Suitability: Tracking Down the 'Beauty Is Beastly' Effect." *Journal of Social Psychology* 150, no. 3 (2010): 301-18.

Jones, Marilyn Y., Andrea J. S. Stanaland, and Betsy D. Gelb. "Beefcake and Cheesecake: Insights for Advertisers." *Journal of Advertising* 27, no. 2 (1998): 33-51.

Jones, Matthew T. "Mediated Exhibitionism: The Naked Body in Performance and Virtual Space." *Sexuality & Culture* 14, no. 4 (2010): 253-69.

Jones, Steve. "The Internet and Its Social Landscape." In *Virtual Culture: Identity and Communication in Cybersociety*, edited by Steve Jones, 7-35. London: Sage Publications, 1997.

Joseph, Branden W. "The Play of Repetition: Andy Warhol's 'Sleep'." *Grey Room*, no. 19 (2005): 22-53.

Juris, Jeffrey S. "Performing Politics: Image, Embodiment, and Affective Solidarity During Anti-Corporate Globalization Protests." *Ethnography* 9, no. 1 (2008): 61-97.

Kahn, Richard, and Douglas Kellner. "New Media and Internet Activism: From the 'Battle of Seattle' to Blogging." *New Media & Society* 6, no. 1 (2004): 87-95.

Kang, Cecilia. "Mothers Rally to Back Breast-Feeding Rights." *Washington Post* November 22, 2006, D1.

Kauffman, L. A. "The Anti-Politics of Identity." *Socialist Review* 20, no. 1 (1990): 67-80.

Keaggy, Diane Toroian. "Cyclists Will Get Naked For ... Does It Matter? St. Louis Will Take Part in World Naked Bike Ride on Saturday, Calling It a Stand against Oil Dependence." *St. Louis Post-Dispatch* July 31, 2008, C4.

Kim, S. W., D. W. Sohn, Y. H. Cho, W. S. Yang, K. U. Lee, R. Juh, K. J. Ahn, Y. A. Chung, S. I. Han, K. H. Lee, C. U. Lee, and J. H. Chae. "Brain Activation by Visual Erotic Stimuli in Healthy Middle Aged Males." *International Journal of Impotence Research* 18, no. 5 (2006): 452-57.

Kirkpatrick, Bill. "'It Beats Rocks and Tear Gas': Streaking and Cultural Politics in the Post-Vietnam Era." *Journal of Popular Culture* 43, no. 5 (2010): 1023-47.

kitchenMage. "Dear PETA, Women Are Animals Too..." http://blog.kitchenmage.com /2007/10/dear-peta-women.html.

Klandermans, Bert. "Mobilization and Participation: Social-Psychological Expansisons of Resource Mobilization Theory." *American Sociological Review* 49, no. 5 (1984): 583-600.

Klandermans, Bert, and Dirk Oegema. "Potentials, Networks, Motivations, and Barriers: Steps Towards Participation in Social Movements." *American Sociological Review* 52, no. 4 (1987): 519-31.

Koukal, D. R. "Here I Stand: Mediated Bodies in Dissent." *MediaTropes* 2, no. 2 (2010): 109-27.

Kratina, Al. "Baring It All to Bring About Change; Nudity Is Optional - but Encouraged - at Today's World Naked Bike Ride." *The Gazette* (Montreal) June 12, 2010, F2.

Krause, Thomas W. "For Safety, Helmets; for Decency, G-Strings." *Tampa Tribune* March 11, 2006, 5.

Krieger, Lisa M. "Nurse in' Planned to Protest Facebook's Ban on Photos of Lactating Breasts." *McClatchy-Tribune News Service* December 26, 2008.

Labre, Magdala Peixoto. "The Brazilian Wax: New Hairlessness Norm for Women?" *Journal Of Communication Inquiry* 26, no. 2 (2002): 113-32.

Lackey, Hilliard. "Running of the Bulls and Running of the Nudes, Intriguing Options." *Mississippi Link* July 6-July 12, 2006, B6.

Laertius, Diogenes. *Lives of Eminent Philosophers*. Translated by Robert Drew Hicks. Vol. 2. Cambridge, MA: Harvard University Press, 1931.

Laner, Mary Riege. "A Rejoinder to Mark Creswell." *Sociological Inquiry* 73, no. 1 (2003): 152-56.

———. "'Sex' Versus 'Gender': A Renewed Plea." *Sociological Inquiry* 70, no. 4 (2000): 462-74.

Lang, Claudia, and Ursula Kuhnle. "Intersexuality and Alternative Gender Categories in Non-Western Cultures." *Hormone Research* 69, no. 4 (2008): 240-50.

Langman, Lauren. "Culture, Identity and Hegemony: The Body in a Global Age." *Current Sociology* 51, no. 3-4 (2003): 223-47.

LaTour, Michael S. "Female Nudity in Print Advertising: An Analysis of Gender Differences in Arousal and Ad Response." *Psychology & Marketing* 7, no. 1 (1990): 65-81.

Leff, Michael C., and Ebony A. Utley. "Instrumental and Constitutive Rhetoric in Martin Luther King Jr.'s 'Letter from Birmingham Jail.'" *Rhetoric & Public Affairs* 7, no. 1 (2004): 37-51.

Lenkowitz, Eric. "Weird but True." *New York Post* June 10, 2007, 23.

Lessig, Lawrence. "The Zones of Cyberspace." *Stanford Law Review* 48, no. 5 (1996): 1403-11.

Levey, Bob. "Breastfeeding in Public: 'Disgusting?'" *Washington Post* June 25, 2001, C10.

Lewis, Claude. "You Can't Grin and Bare It in Public; Americans' Breast Obsession Doesn't Extend to Breast-Feeding." *Columbia Daily Tribune* (Columbia, MO) June 19, 2007, http://archive.columbiatribune.com/2007/jun/20070619comm011.asp.

Li, Chi-Ying, Oliver Kayes, Phillip D. Kell, Nim Christopher, Suks Minhas, and David J. Ralph. "Penile Suspensory Ligament Division for Penile Augmentation: Indications and Results." *European Urology* 49, no. 4 (2006): 729-33.

Libertson, Joseph. *Proximity, Levinas, Blanchot, Bataille, and Communication*. The Hague: Martinus Nijhoff, 1982.

Lindberg, Laura Duberstein. "Women's Decisions About Breastfeeding and Maternal Employment." *Journal of Marriage and the Family* 58, no. 1 (1996): 239-51.

Ling, Peter. "Sex and the Automobile in the Jazz Age." *History Today* 39, no. 11 (1989): 18-24.

Linz, Daniel, Eva Blumenthal, Edward Donnerstein, Dale Kunkel, Bradley J. Shafer, and Allen Lichtenstein. "Testing Legal Assumptions Regarding the Effects of Dancer Nudity and Proximity to Patron on Erotic Expression." *Law and Human Behavior* 24, no. 5 (2000): 507-33.

Linzey, Andrew. "Good Causes Do Not Need Exaggeration." *The Animals' Agenda* 20, no. 1 (2000): 24-25.

Love, Brenda. "Lactaphilia." In *The Encyclopedia of Unusual Sex Practices*, edited by Brenda Love, 151-52. London: Greenwich Editions, 1992.

Lunceford, Brett. "Clothes Make the Person? Performing Gender through Fashion." *Communication Teacher* 24, no. 2 (2010): 63-68.

———. "PETA and the Rhetoric of Nude Protest." In *Arguments About Animal Ethics*, edited by Greg Goodale and Jason Edward Black, 97-112. Lanham, MD: Lexington Books, 2010.

———. "Smeared Makeup and Stiletto Heels: Clothing, Sexuality, and the Walk of Shame." In *College Sex: Philosophy for Everyone: Philosophers with Benefits*, edited by Robert Stewart and Michael Bruce, 51-60. Hoboken, NJ: Wiley-Blackwell, 2010.

———. "The Science of Orality: Implications for Rhetorical Theory." *Review of Communication* 7, no. 1 (2007): 83-102.

———. "The Walk of Shame: A Normative Description." *ETC: A Review of General Semantics* 65, no. 4 (2008): 319-29.

MacDonald, Shauna M. "Leaky Performances: The Transformative Potential of Menstrual Leaks." *Women's Studies in Communication* 30, no. 3 (2007): 340-57.

Marshall, W. L., Krista Payne, H. E. Barbaree, and A. Eccles. "Exhibitionists: Sexual Preferences for Exposing." *Behaviour Research and Therapy* 29, no. 1 (1991): 37-40.

Martin, Shelby. "Cyclists in Bike Parade to Air Views in Ride of Protest." *Register-Guard* (Eugene, OR) March 14, 2009, A1.

Masters, William H., and Virginia E. Johnson. *Human Sexual Inadequacy*. Boston, MA: Little, Brown and Company, 1970.

Matheson, Peter. "The Hammer, the Sickle and the Rainbow: Sixteenth-Century Radicalism and Its Interpreters." *Theology* 93, no. 751 (1990): 20-26.

Matthews, Melvyn. "'Who Told You That You Were Naked?' A Metaphor for a Christian Theology of Atonement." *Theology* 93, no. 753 (1990): 212-20.

McCarthy, John D., and Mayer N. Zald. "Resource Mobilization and Social Movements: A Partial Theory." *American Journal of Sociology* 82, no. 6 (1977): 1212-41.

———. "The Trend of Social Movements in America: Professionalization and Resource Mobilization." In *Social Movements in an Organizational Society: Collected Essays*, edited by Mayer N. Zald and John D. McCarthy, 337-91. New Brunswick, NJ: Transaction Books, 1987.

McCartney, Jenny. "Why's Facebook So Het up About Breast-Feeding?" *Sunday Telegraph* (London) January 4, 2009, 23.

McCombs, Maxwell E., and Donald L. Shaw. "The Agenda-Setting Function of Mass Media." *Public Opinion Quarterly* 36, no. 2 (1972): 176-87.

McGee, Michael C. "In Search of 'The People': A Rhetorical Alternative." *Quarterly Journal of Speech* 61, no. 3 (1975): 235-49.

McGee, Michael Calvin. "'Social Movement': Phenomenon or Meaning?" *Central States Speech Journal* 31 (1980): 233-44.

———. "Text, Context, and the Fragmentation of Contemporary Culture." *Western Journal of Communication* 54, no. 3 (1990): 274-89.

McKinlay, Judith E. "Bothering to Enter the Garden of Eden Once Again." *Feminist Theology* 19, no. 2 (2011): 143-53.

McLuhan, Marshall. *Understanding Media: The Extensions of Man*. Cambridge, MA: MIT Press, 1994.

McLuhan, Marshall, Quentin Fiore, and Jerome Agel. *The Medium Is the Massage: An Inventory of Effects*. San Francisco, CA: HardWired, 1996.

McNair, Brian. *Striptease Culture: Sex, Media and the Democratization of Desire*. London: Routledge, 2002.

Melucci, Alberto. *Challenging Codes: Collective Action in the Information Age*. Cambridge: Cambridge University Press, 1996.

Messner, Matthias, Marc-André Reinhard, and Siegfried Ludwig Sporer. "Compliance through Direct Persuasive Appeals: The Moderating Role of Communicator's Attractiveness in Interpersonal Persuasion." *Social Influence* 3, no. 2 (2008): 67-83.

Meston, Cindy M., and Boris B. Gorzalka. "Differential Effects of Sympathetic Activation on Sexual Arousal in Sexually Dysfunctional and Functional Women." *Journal of Abnormal Psychology* 105, no. 4 (1996): 582-91.

Mesure, Susie. "Facebook Stirs up Another Row with Nursing Mothers." *Independent on Sunday* (London) January 9, 2011, 12.

Meyer, David S. *The Politics of Protest: Social Movements in America*. New York: Oxford University Press, 2007.

Michaud, Leonard J. "Decent Exposure: An Inquiry into the Constitutionality of State Statutes Proscribing Expressive Public Nudity Comment." *Loyola Law Review* 22 (1975): 1018-35.

Mika, Marie. "Framing the Issue: Religion, Secular Ethics and the Case of Animal Rights Mobilization." *Social Forces* 85, no. 2 (2006): 915-41.

"M.I.L.C. Virtual Nurse-In." Facebook, www.facebook.com/event.php?eid =102988089741.

Milstead, Virginia F. "Forbidding Female Toplessness: Why Real Difference Jurisprudence Lacks Support and What Can Be Done About It." *University of Toledo Law Review* 36, no. 2 (2005): 273-320.

Mintz, Jessica. "Nursing Moms Mad over Facebook Policy." *Charleston Daily Mail* January 2, 2009, A8.

Mishra, Raja. "Nursing Mother's Protest Grows; Nursing Mother's Complaint Spurs Plans to Target Delta Today Organizers Target Delta Today." *Boston Globe* November 21, 2006, A1.

Mitchell, Timothy J. "Bullfighting: The Ritual Origin of Scholarly Myths." *Journal of American Folklore* 99 (1986): 394-414.

Mohammad, Susan. "Protesters Grin and Bare It Despite Chilly Conditions." *Ottawa Citizen* June 11, 2006, A11.

Møller, Anders Pape, Manuel Soler, and Randy Thornhill. "Breast Asymmetry, Sexual Selection, and Human Reproductive Success." *Ethology and Sociobiology* 16, no. 3 (1995): 207-19.

Moloney, Paul. "A Wheely Cheeky Protest; World Naked Bike Ride Cyclists Doff Duds to Promote 2-Wheeled Solution to Air Pollution." *Toronto Star* June 10, 2007, A2.

"Moms Stage 'Nurse-in' at Nation's Airports." *Deseret News* (Salt Lake City, UT) November 22, 2006, A6.

Moreland, Jay, and Jennifer Coombs. "Promoting and Supporting Breast-Feeding." *American Family Physician* 61, no. 7 (2000): 2093-100.

Moriber, Danielle. "A Right to Bare All? Female Public Toplessness and Dealing with the Laws That Prohibit." *Cardozo Public Law, Policy, and Ethics Journal* 8, no. 2 (2010): 453-84.

Morita, Jennifer K. "Moms Fight for Right to Nurse in Public ; Despite Laws, Society Still Squeamish." *The Record* (Bergen County, NJ) August 18, 2006, A25.

Morland, Iain. "'The Glans Opens Like a Book': Writing and Reading the Intersexed Body." *Continuum: Journal of Media & Cultural Studies* 19, no. 3 (2005): 335-48.

Morris, Dick. "The Future of Political Campaigning: The American Example." *Journal of Public Affairs* 3, no. 1 (2003): 14-20.

"Mother's Outrage Prompts Bill on the Right to Nurse in Public." *New York Times* February 16, 1997, 47.

Motofel, Ion G., and David L. Rowland. "The Physiological Basis of Human Sexual Arousal: Neuroendocrine Sexual Asymmetry." *International Journal of Andrology* 28, no. 2 (2005): 78-87.

Mulvey, Laura. "Visual Pleasure and Narrative Cinema." *Screen* 16, no. 3 (1975): 6-18.

Murray, Erin K., Sue Ricketts, and Jennifer Dellaport. "Hospital Practices That Increase Breastfeeding Duration: Results from a Population-Based Study." *Birth: Issues in Perinatal Care* 34, no. 3 (2007): 202-11.

Negroponte, Nicholas. *Being Digital*. New York: Knopf, 1995.

Nelson, Michelle R., and Hye-Jin Paek. "Nudity of Female and Male Models in Primetime TV Advertising across Seven Countries." *International Journal of Advertising* 27, no. 5 (2008): 715-44.

Nelson, Thomas E., and Zoe M. Oxley. "Issue Framing Effects on Belief Importance and Opinion." *Journal of Politics* 61, no. 4 (1999): 1040-67.

NHTSA's National Center for Statistics and Analysis. "Traffic Safety Facts 2009 Data: Bicyclists and Other Cyclists." Washington, DC: National Highway Traffic Safety Administration.

Noguchi, Sharon. "Online Protest of Facebook Ban on Breast-Feeding Photos Draws Tens of Thousands." *McClatchy-Tribune News Service* December 27, 2008.

Norton, Laura H. "Neutering the Transgendered: Human Rights and Japan's Law No. 111." *Georgetown Journal of Gender & the Law* 7, no. 2 (2006): 187-216.

Nosko, Amanda, Eileen Wood, and Serge Desmarais. "Unsolicited Online Sexual Material: What Affects Our Attitudes and Likelihood to Search for More?" *Canadian Journal of Human Sexuality* 16, no. 1/2 (2007): 1-10.

Oliver, Pamela E., and Daniel J. Myers. "How Events Enter the Public Sphere: Conflict, Location, and Sponsorship in Local Newspaper Coverage of Public Events." *American Journal of Sociology* 105, no. 1 (1999): 38-87.

O'Malley, Nick, and Jonathan Pearlman. "Raelian around the GM Flag." *Sydney Morning Herald* August 8, 2003, 16.

Oremusk, Will. "Nursing Moms Take Aim at Facebook over Breastfeeding Censorship." *Oakland Tribune* December 17, 2008.

Orleck, Annelise. "'We Are That Mythical Thing Called the Public': Militant Housewives During the Great Depression." *Feminist Studies* 19, no. 1 (1993): 147-72.

Owens, Lynn, and L. Kendall Palmer. "Making the News: Anarchist Counter-Public Relations on the World Wide Web." *Critical Studies in Media Communication* 20, no. 4 (2003): 335-61.

Pace, Julie. "Protestors Shun Bike Shorts to Bare Views on Oil, Cars." *Tampa Tribune* March 13, 2006, 6.

Pace, Lesli K. "Image Events and PETA's Anti Fur Campaign." *Women and Language* 28, no. 2 (2005): 33-41.

Padilla, Andrew. "Mothers Show for 'Nurse-in' Protest." *Albuquerque Journal* October 6, 1999, B1.

Paek, Hye-Jin, and Michelle R. Nelson. "A Cross-Cultural and Cross-Media Comparison of Female Nudity in Advertising." *Journal of Promotion Management* 13, no. 1/2 (2007): 145-67.

Papacharissi, Zizi. "The Virtual Sphere: The Internet as a Public Sphere." *New Media & Society* 4, no. 1 (2002): 9-27.

Payne, Elizabeth. "Still Squeamish About Breastfeeding." *Ottawa Citizen* August 19, 2008, A10.

Payne, Jan, Barbara Radcliffe, Emma Blank, Elizabeth Churchill, Nadia Hassan, Elizabeth Cox, and Helen Porteous. "Breastfeeding: The Neglected Guideline for Future Dietitian-Nutritionists?" *Nutrition & Dietetics* 64, no. 2 (2007): 93-98.

Pearson, Jo. "Inappropriate Sexuality? Sex Magic, S/M and Wicca (or 'Whipping Harry Potter's Arse!')." *Theology & Sexuality* 11, no. 2 (2005): 31-42.

Peirce, Charles S., Charles Hartshorne, and Paul Weiss, eds. *Collected Papers*. 6 vols. Cambridge, MA: Belknap Press of Harvard University Press, 1960.

People for the Ethical Treatment of Animals. "PETA Holds Nude Conferences in New Plan to Save Bulls." www.runningofthenudes.com/nude_conferences.asp.

———. "PETA's 2007 State of the Union Undress." features.peta.org/stateoftheunion2007/f-stateoftheunion.asp.

———. "Running of the Nudes: Out with the Old-in with the Nude!" www.runningofthenudes.com.

———. "Running of the Nudes: Out with the Old-in with the Nude! // Next Year's Race // Look Who's Running." www.runningofthenudes.com/look_whos_running.asp.

Perloff, Richard M. "The Third Person Effect in Media Effects." In *Media Effects: Advances in Theory and Research*, edited by Jennings Bryant and Dolf Zillmann, 489-506. Mahwah, NJ: Lawrence Elbaum Associates, 2002.

Peterson, Abby. *Contemporary Political Protest : Essays on Political Militancy*. Aldershot, England: Ashgate, 2001.

Peterson, Valerie V. "The Rhetorical Criticism of Visual Elements: An Alternative to Foss's Schema." *Southern Communication Journal* 67, no. 1 (2001): 19-32.

Phelan, Sharon T. "Fads and Fashions: The Price Women Pay." *Primary Care Update for OB/GYNS* 9, no. 4 (2002): 138-43.

Phillips, Barbara J., and Edward F. McQuarrie. "Beyond Visual Metaphor: A New Typology of Visual Rhetoric in Advertising." *Marketing Theory* 4 (2004): 113-36.

Pitts, Victoria L. "'Reclaiming' the Female Body: Embodied Identity Work, Resistance and the Grotesque." *Body & Society* 4, no. 3 (1998): 67-84.

Plato. "Phaedrus." In *The Collected Dialogues of Plato, Including the Letters*, edited by Edith Hamilton and Huntington Cairns, 475-525. Princeton, NJ: Princeton University Press, 1961.

Pomeroy, Sarah B. *Goddesses, Whores, Wives, and Slaves: Women in Classical Antiquity*. New York: Schocken Books, 1975.

Postman, Neil. *Technopoly: The Surrender of Culture to Technology*. New York: Vintage Books, 1993.

Poulakos, John. "Toward a Sophistic Definition of Rhetoric." *Philosophy & Rhetoric* 16, no. 1 (1983): 35-48.

Pratt, John. "Pornography and Everyday Life." *Theory Culture & Society* 3, no. 1 (1986): 65-78.

"Protesters Take Part in 'Nurse-Ins' at Airports." *Capital* (Annapolis, MD) November 21, 2006, A4.

Pundurs, Helen. "Public Exposure of the Female Breast: Obscene and Immoral or Free and Equal." *In the Public Interest* 14 (1994-1995): 1-38.

Putnam, Robert D. *Bowling Alone: The Collapse and Revival of American Community*. New York: Simon & Schuster, 2000.

Radford, Bill. "Bullfight Protester Goes Bare; Woman with Local Ties Works for PETA." *The Gazette* (Colorado Springs) June 25, 2005, Life3.

Ramirez, Anthony. "War Protester in Village Takes Her Cue from Lady Godiva." *New York Times* August 10, 2005, B4.

Rathke, Lisa. "Teens Test Naked City's Limits; Town Asks: Should There Be a Law?" *Washington Post* September 24, 2006, A8.

Razaq, Rashid. "Naked Cyclists to Stage London Demo." *Evening Standard* (London) June 5, 2006, 19.

"Readers Divided on the Issue." *Toronto Star* January 30, 2007, E5.

Redmon, David. "Playful Deviance as an Urban Leisure Activity: Secret Selves, Self-Validation, and Entertaining Performances." *Deviant Behavior* 24, no. 1 (2003): 27-51.

Reichert, Tom. "Sex in Advertising Research: A Review of Content, Effects, and Functions of Sexual Information in Consumer Advertising." *Annual Review of Sex Research* 13 (2002): 241-73.

Reinert, Sue. "Nursing Moms Defend Practice; Protest Comment by Barbara Walters." *Patriot Ledger* (Quincy, MA) June 13, 2005, 1.

Reingen, Peter H., and Jerome B. Kernan. "Social Perception and Interpersonal Influence: Some Consequences of the Physical Attractiveness Stereotype in a Personal Selling Setting." *Journal of Consumer Psychology* 2, no. 1 (1993): 25-38.

Reinhard, Marc-André, Matthias Messner, and Siegfried Ludwig Sporer. "Explicit Persuasive Intent and Its Impact on Success at Persuasion—The Determining Roles of Attractiveness and Likeableness." *Journal of Consumer Psychology* 16, no. 3 (2006): 249-59.

Richards, Gwendolyn. "Naked Protesters Plan Anti-War Display Whatever the Weather." *Vancouver Sun* February 22, 2003, B4.

"Right to Breast-Feed Needs Protection." *Times Union* (Albany, NY) December 10, 1996, A14.

Roberts, Roxanne. "Do Me a Favor, Keep a Lid on Your Double Latte." *The Washington Post* August 11, 2004, C1.

Roberts, S. Craig, Emily J. Miner, and Todd K. Shackelford. "The Future of an Applied Evolutionary Psychology for Human Partnerships." *Review of General Psychology* 14, no. 4 (2010): 318-29.

Romboy, Dennis. "Moms Stage 'Nurse-in' at Burger King." *Deseret News* (Salt Lake City, UT) November 11, 2003, A1.

Roseingrave, Louise. "Wheely Brave." *Daily Mirror* (London) July 20, 2010, 3.

Rosenberg, Kenneth D., Carissa A. Eastham, Laurin J. Kasehagen, and Alfredo P. Sandoval. "Marketing Infant Formula through Hospitals: The Impact of Commercial Hospital Discharge Packs on Breastfeeding." *American Journal of Public Health* 98, no. 2 (2008): 290-95.

Rosenblum, Gail. "Feuding over PDL*; *Public Displays of Lactation; A Breast-Feeding Controversy Has Reignited the Battle among Mothers over What Is Proper While Nursing Their Babies." *Star Tribune* (Minneapolis, MN) January 31, 2009, E1.

Rubin, Bonnie Miller. "Laws Aim to Protect Moms Who Breast-Feed." *Chicago Tribune* February 17, 2003, 1.

Saco, Diana. *Cybering Democracy: Public Space and the Internet*. Minneapolis: University of Minnesota Press, 2002.

Saillant, Catherine. "Ojai's Pastie Lady Takes Her Activism Elsewhere; She Heads to Oregon, Looking for Less Hassle and Greener Mind-Sets." *Los Angeles Times* May 11, 2008, B3.

———. "She's Got Ojai's Attention; an Activist Who Bikes through Town in Only Pasties and a G-String Has Supporters and Critics. An Easter Incident Could Lead to Charges." *Los Angeles Times* April 26, 2008, B1.

Saker, Anne. "Cyclist Was Ready to Lay Bare Case—Until It Was Dismissed / Minister Was Arrested on World Naked Bike Ride, but Doesn't Need His Day in Court." *Houston Chronicle* July 1, 2007, 6.

Scammell, Margaret. "The Internet and Civic Engagement: The Age of the Citizen-Consumer." *Political Communication* 17, no. 4 (2000): 351-55.

Schiff, Lauren. "Breastfeeding Makes for Better Health." *Mount Sinai Journal of Medicine* 73, no. 2 (2006): 571-72.

Schupp, Harald T., Bruce N. Cuthbert, Margaret M. Bradley, Charles H. Hilman, Alfons O. Hamm, and Peter J. Lang. "Brain Processes in Emotional Perception: Motivated Attention." *Cognition & Emotion* 18, no. 5 (2004): 593-611.

Schwartau, Winn. *Cybershock: Surviving Hackers, Phreakers, Identity Thieves, Internet Terrorists and Weapons of Mass Destruction*. New York: Thunder's Mouth Press, 2000.

Schwartz, Jeremy. "Cheeky Cyclists to Cruise Austin in Clothing-Free Ride." *Austin American Statesman* June 11, 2004, A1.

Scorolli, C., S. Ghirlanda, M. Enquist, S. Zattoni, and E. A. Jannini. "Relative Prevalence of Different Fetishes." *International Journal of Impotence Research* 19, no. 4 (2007): 432-37.

Scudder, Joseph N., and Carol Bishop Mills. "The Credibility of Shock Advocacy: Animal Rights Attack Messages." *Public Relations Review* 35, no. 2 (2009): 162-64.

Segal, Charles P. "Gorgias and the Psychology of the Logos." *Harvard Studies in Classical Philology* 66 (1962): 99-155.

Seigneur, Cornelia M. "Moms Nurse in Protest at Water Park." *Oregonian* (Portland, OR) February 21, 1998, B6.

Sengupta, Jaideep, and Darren W. Dahl. "Gender-Related Reactions to Gratuitous Sex Appeals in Advertising." *Journal of Consumer Psychology* 18, no. 1 (2008): 62-78.

Sennett, Richard. *The Fall of Public Man*. New York: W.W. Norton, 1996.

Severn, Jessica, George E. Belch, and Michael A. Belch. "The Effects of Sexual and Non-Sexual Advertising Appeals and Information Level on Cognitive Processing and Communication Effectiveness." *Journal of Advertising* 19, no. 1 (1990): 14-22.

Shaw, Sylvie. "Reclaiming the Ecoerotic: Celebrating the Body and the Earth." *Ecotheology: Journal of Religion, Nature & the Environment* 8, no. 1 (2003): 85-99.

Sherman, Claire, and Pascale Quester. "The Influence of Product/Nudity Congruence on Advertising Effectiveness." *Journal of Promotion Management* 11, no. 2/3 (2005): 61-89.

Shildrick, Margrit. *Leaky Bodies and Boundaries: Feminism, Postmodernism and (Bio)Ethics*. London: Routledge, 1997.

Simons, Herbert W. "Requirements, Problems, and Strategies: A Theory of Persuasion for Social Movements." *Quarterly Journal of Speech* 56, no. 1 (1970): 1-11.

Simonson, Peter. "Social Noise and Segmented Rhythms: News, Entertainment, and Celebrity in the Crusade for Animal Rights." *Communication Review* 4, no. 3 (2001): 399-420.

Simpson, Penny M., Steve Horton, and Gene Brown. "Male Nudity in Advertisements: A Modified Replication and Extension of Gender and Product Effects." *Journal of the Academy of Marketing Science* 24, no. 3 (1996): 257-62.

Singer, Ross. "Anti-Corporate Argument and the Spectacle of the Grotesque Rhetorical Body in Super Size Me." *Critical Studies in Media Communication* 28, no. 2 (2011): 135-52.

Singh, Devendra, and Robert K. Young. "Body Weight, Waist-to-Hip Ratio, Breasts, and Hips: Role in Judgments of Female Attractiveness and Desirability for Relationships." *Ethology and Sociobiology* 16, no. 6 (1995): 483-507.

Smelser, Neil J. *Theory of Collective Behavior*. New York: Free Press, 1962.

Smith, Elizabeth. "Breastfeeding for a Healthy Lifestyle: Is Our Education Working?" *International Journal of Childbirth Education* 22, no. 1 (2007): 30-33.

———. "Is Breast Still Best? How Are Environmental Toxins Affecting Breast Milk?" *International Journal of Childbirth Education* 23, no. 2 (2008): 12-14.

Smith, Jackie, John D. McCarthy, Clark McPhail, and Boguslaw Augustyn. "From Protest to Agenda Building: Description Bias in Media Coverage of Protest Events in Washington, D.C." *Social Forces* 79, no. 4 (2001): 1397-423.

Smith, Moira. "Humor, Unlaughter, and Boundary Maintenance." *Journal of American Folklore* 122, no. 484 (2009): 148-71.

Sobchack, Vivian. "Beating the Meat/Surviving the Text, or How to Get out of This Century Alive." In *Cyberspace/Cyberbodies/Cyberpunk: Cultures of Technological Embodiment*, edited by Mike Featherstone and Roger Burrows, 205-14. London: Sage, 1995.

———. *Carnal Thoughts: Embodiment and Moving Image Culture*. Berkeley: University of California Press, 2004.

Soley, Lawrence C., and Leonard N. Reid. "Taking It Off: Are Models in Magazine Ads Wearing Less?" *Journalism Quarterly* 65, no. 4 (1988): 960-66.

"Some Private Photos Ought to Stay That Way." *Herald* (Everett, WA) January 5, 2009.

Spear, Hila J. "Breast Is Best and Culture Wars." *International Journal of Childbirth Education* 22, no. 1 (2007): 34-35.

Speidel, Michael P. "Berserks: A History of Indo-European 'Mad Warriors'." *Journal of World History* 13, no. 2 (2002): 253-90.

Spiering, Mark, Walter Everaerd, and Erick Janssen. "Priming the Sexual System: Implicit Versus Explicit Activation." *Journal of Sex Research* 40, no. 2 (2003): 134-45.

Spinoza, Baruch. *The Ethics; Treatise on the Emendation of the Intellect; Selected Letters*. Edited by Seymour Feldman. Translated by Samuel Shirley. 2nd ed. Indianapolis, IN: Hackett, 1992.

Spurles, Patricia Kelly, and Jessica Babineau. "A Qualitative Study of Attitudes toward Public Breastfeeding among Young Canadian Men and Women." *Journal of Human Lactation* 27, no. 2 (2011): 131-37.

Stack, Liam, and David D. Kirkpatrick. "Nude Blogger Riles Egyptians of All Stripes." *New York Times* November 18, 2011 4.

Staggenborg, Suzanne. "The Consequences of Professionalization and Formalization in the Pro-Choice Movement." *American Sociological Review* 53, no. 4 (1988): 585-605.

Stanton, Susan L. *Being Naked: Attitudes Toward Nudity Through the Ages*. St. Clair Shores, MI: Ablaze Press, 2001.

Stearns, Cindy A. "Breastfeeding and the Good Maternal Body." *Gender and Society* 13, no. 3 (1999): 308-25.

Stephens, William N. "A Cross-Cultural Study of Modesty." *Cross-Cultural Research* 7, no. 1 (1972): 1-28.

Sterling, Andrew. "Director Offers Version of the Hole Truth." *Columbus Dispatch* May 20, 2005, E12.

Stevens, Phillips, Jr. "Women's Aggressive Use of Genital Power in Africa." *Transcultural Psychiatry* 43, no. 4 (2006): 592-99.

Stewart, Charles J. "Championing the Rights of Others and Challenging Evil: The Ego Function in the Rhetoric of Other-Directed Social Movements." *Southern Communication Journal* 64, no. 2 (1999): 91-105.

Stilts, Josh. "Breastfeeding Marlboro Mom Wants Answers from Facebook." *Bennington Banner* (Bennington, VT) February 1, 2011.

Stinson, Scott. "Barbara Walters Target of 'Nurse-in': 'Anti-Breastfeeding.'" *National Post* (Don Mills, ON) June 7, 2005, A3.

Stovall, James Glen. *Writing for the Mass Media*. 5th ed. Boston: Allyn and Bacon, 2002.

Sullivan, Laura L. "Cyberbabes: (Self-) Representation of Women and the Virtual Male Gaze." *Computers and Composition* 14, no. 2 (1997): 189-204.

"Superbeauty.Org—Beauty Will Save the World!" SuperBeauty.Org, www.superbeauty.org.

Sutton, Barbara. "*Poner El Cuerpo*: Women's Embodiment and Political Resistance in Argentina." *Latin American Politics & Society* 49, no. 3 (2007): 129-62.

Tait, James. "Buff Riders Saddle the Blame." *Daily Mirror* (London) June 12, 2004, 27.

"Take It Off." *Chicago Tribune* May 15, 2009, 36.

"Take Off Your Car and Join the Nude Cyclists." *Ottawa Citizen* December 26, 2007, A17.

Tamas, Sophie. "Sketchy Rendering: Seeing an Other." *Qualitative Inquiry* 15, no. 3 (2009): 607-17.

Tapper, Kathy. "Asserting Their Right to Feed Their Children; 266 Mothers Breast-Feed at the Cove." *South Bend Tribune* (South Bend, IN) August 3, 2003, 1.

Tate, Helen. "The Ideological Effects of a Failed Constitutive Rhetoric: The Co-Option of the Rhetoric of White Lesbian Feminism." *Women's Studies in Communication* 28, no. 1 (2005): 1-31.

Taylor, Verta. "Mobilizing for Change in a Social Movement Society." *Contemporary Sociology* 29, no. 1 (2000): 219-30.

Terranova, Tiziana. "Communication Beyond Meaning: On the Cultural Politics of Information." *Social Text* 22, no. 3 (2004): 51-73.

"The Shape of a Mother." The Shape of a Mother, http://theshapeofamother.com/.

Thomas, Douglas. *Hacker Culture*. Minneapolis: University of Minnesota Press, 2002.

Tiggemann, Marika, Yolanda Martins, and Libby Churchett. "Beyond Muscles: Unexplored Parts of Men's Body Image." *Journal of Health Psychology* 13, no. 8 (2008): 1163-72.

Touraine, Alain. *What Is Democracy?* Translated by David Macey. Boulder, CO: Westview Press, 1997.

Trapasso, Clare, and Jotham Sederstrom. "Bikers Strip for Rally vs. Cars and Oil." *New York Daily News* June 14, 2009, 2.

Trout, Becky. "Striptease to Save Trees." *American Forests* 107, no. 3 (2001): 19.

Tsoffar, Ruth. "The Body as Storyteller: Karaite Women's Experience of Blood and Milk." *Journal of American Folklore* 117, no. 463 (2004): 3-21.

Turkle, Sherry. *Life on the Screen: Identity in the Age of the Internet*. New York: Simon & Schuster, 1995.

Turner, Terisa E., and Leigh S. Brownhill. "Why Women Are at War with Chevron: Nigerian Subsistence Struggles against the International Oil Industry." *Journal of Asian and African Studies* 39, no. 1-2 (2004): 63-93.

Turton-Turner, Pamela. "The Role of Ridicule in Naked Charity Calendars." *Continuum: Journal of Media & Cultural Studies* 21, no. 3 (2007): 419-32.

Uhlin, Graig. "TV, Time, and the Films of Andy Warhol." *Cinema Journal* 49, no. 3 (2010): 1-23.

"Up to 200 Naked Cyclists Expected on City's Streets." *Irish Examiner* July 19, 2010.

Van Esterik, Penny, and Ted Greiner. "Breastfeeding and Women's Work: Constraints and Opportunities." *Studies in Family Planning* 12, no. 4 (1981): 184-97.

Vardi, Yoram, Yaron Harshai, Tamir Gil, and Ilan Gruenwald. "A Critical Analysis of Penile Enhancement Procedures for Patients with Normal Penile Size: Surgical Techniques, Success, and Complications." *European Urology* 54, no. 5 (2008): 1042-50.

Velleman, James David. "The Genesis of Shame." *Philosophy & Public Affairs* 30, no. 1 (2001): 27-52.

Verdi, Laura. "The Symbolic Body and the Rhetoric of Power." *Social Analysis* 54, no. 2 (2010): 99-115.

Walters, Suzanna Danuta. "Caged Heat: The (R)Evolution of Women-in-Prison Films." In *Reel Knockouts: Violent Women in the Movies*, edited by Martha McCaughey and Neal King, 106-23. Austin, TX: University of Texas Press, 2001.

Warner, Michael. *Publics and Counterpublics*. New York: Zone Books, 2002.

Watzlawick, Paul, Janet Beavin Bavelas, and Don D. Jackson. *Pragmatics of Human Communication: A Study of Interactional Patterns, Pathologies, and Paradoxes*. New York: W. W. Norton & Company, 1967.

Wentzell, Emily, and Jorge Salmerón. "You'll 'Get Viagraed': Mexican Men's Preference for Alternative Erectile Dysfunction Treatment." *Social Science & Medicine* 68, no. 10 (2009): 1759-65.

Westdahl, Claire, and Sallie Page-Goertz. "Promotion of Breastfeeding—Beyond the Benefits." *International Journal of Childbirth Education* 21, no. 4 (2006): 8-16.

Whittle, Stephen. "The Opposite of Sex Is Politics—The UK Gender Recognition Act and Why It Is Not Perfect, Just Like You and Me." *Journal of Gender Studies* 15, no. 3 (2006): 267-71.

Whitty, Monica Therese. "Cyber-Flirting: Playing at Love on the Internet." *Theory & Psychology* 13, no. 3 (2003): 339-57.

Windt, Theodore Otto, Jr. "The Diatribe: Last Resort for Protest." *Quarterly Journal of Speech* 58 (1972): 1-14.

Winterburn, S. "Breastfeeding Support Plans: An Evidenced-Based Approach." *Primary Health Care* 15, no. 4 (2005): 36-39.

"With Friends Like These, Monsanto Needs No Enemies." USA Today, www.usatoday .com/tech/news/techpolicy/2003-08-06-raelians-biotech_x.htm.

Wolf, Jacqueline H. "Low Breastfeeding Rates and Public Health in the United States." *American Journal of Public Health* 93, no. 12 (2003): 2000-10.

Wolf, Joan B. "Is Breast Really Best? Risk and Total Motherhood in the National Breastfeeding Awareness Campaign." *Journal of Health Politics, Policy & Law* 32, no. 4 (2007): 595-636.

World Naked Bike Ride. "Organizing a Ride." wiki.worldnakedbikeride.org/index.php ?title=Organizing_a_ride.

———. "World Naked Bike Ride (WNBR) - Naked Bicycle People Power! Stop Indecent Exposure to Vehicle Emissions!" www .worldnakedbikeride.org.

"World Naked Bike Ride." Wikipedia, en.wikipedia.org/wiki/World_Naked_Bike _Ride.

Wray, Stefan. "On Electrical Civil Disobedience." *Peace Review* 11, no. 1 (1999): 107-11.

Wynne, Sharon Kennedy. "Nursing Moms to Protest Online." *St. Petersburg Times* December 26, 2008, B1.

———. "Nursing Mothers Take on Facebook." *Seattle Post-Intelligencer* December 27, 2008, A10.

Young, Diony. "Violating 'the Code': Breastfeeding, Ethics, and Choices." *Birth: Issues in Perinatal Care* 28, no. 2 (2001): 77-78.

Young, Iris Marion. *On Female Body Experience: "Throwing Like a Girl" and Other Essays*. New York: Oxford University Press, 2005.

Zaretsky, Eli. "Identity Theory, Identity Politics: Psychoanalysis, Marxism, Post-Structuralism." In *Social Theory and the Politics of Identity*, edited by Craig J. Calhoun, 198-215. Oxford: Blackwell, 1994.

Zerbisias, Antonia. "Are These Obscene?" *Toronto Star* April 13, 2010, A3.

Index

About the Author

Brett Lunceford (PhD, Pennsylvania State University) is an assistant professor in the Department of Communication at the University of South Alabama. His work has appeared in several edited volumes, as well as in *American Communication Journal, Communication Teacher, ETC: A Review of General Semantics, Explorations in Media Ecology, Media History Monographs, Northwestern Journal of Technology and Intellectual Property, Review of Communication,* and *Theology and Sexuality*. He lives in Mobile, Alabama with his wife, Rebecca, and two sons, Adam and Soren.

www.ingramcontent.com/pod-product-compliance
Lightning Source LLC
LaVergne TN
LVHW101448200926
848483LV00001B/88
* 9 7 8 0 7 3 9 1 6 7 0 9 0 *